International Harvester Tractor: Identification Guide

1939-1985

By Kenneth Updike

1831 Press
December 1, 2022

Dedication

This book is dedicated to everyone who has ever owned a McCormick or Farmall Tractor.

Acknowlegements

The author would like to thank the following for thier help in the construction of this book: Betty Rogers, John Henigan, Robert Zarse, Darrell Darst, Johnson Tractor, Sarah Tomac and myself.

All Serial Number and Production Data in this book has originated from official IH records

IBSN: 9781952265075
Printed and distributed by 1831 Press

Contents

Page - Tractor Model

Page - Tractor Model

Page - Tractor Model

Page - Tractor Model

TRACTOR MODEL   Farmall Cub
PLACE OF MANUFACTURE Louisville
RATED HORSEPOWER   BELT   10
NEBRASKA TEST #   386
ENGINE   C60
BORE & STROKE   2-5/8" x 2-3/4"
AVERAGE SHIPPING WEIGHT   1,400 lbs
TOTAL UNITS BUILT   224,203

TRACTOR SERIAL NUMBER DATA

Serial Numbers;	Year built;	Total units
501 to 11347	Built in 1947	10,847
11348 to 57830	Built in 1948	46,483
57831 to 99535	Built in 1949	41,705
99536 to 121453	Built in 1950	21,918
121454 to 144454	Built in 1951	23,001
144455 to 162283	Built in1952	17,829
162284 to 179411	Built in 1953	17,128
179412 to 186440	Built in 1954	7,029
186441 to 193657	Built in 1955	7,217
193658 to 198230	Built in 1956	4,573
198231 to 204388	Built in 1957	6,158
204389 to 211440	Built in 1958	7,052
211441 to 214973	Built in 1959	3,533
214974 to 217381	Built in 1960	2,408
217382 to 220037	Built in 1961	2,656
220038 to 221382	Built in 1962	1,344
221383 to 223452	Built in 1963	2,070
223453 to 224703	Built in 1964	1,251

TRACTOR MODEL   International Cub
PLACE OF MANUFACTURE Louisville
RATED HORSEPOWER   BELT   10
NEBRASKA TEST #   386
ENGINE   C60
BORE & STROKE   2-5/8" x 2-3/4"
AVERAGE SHIPPING WEIGHT   1,400 lbs
TOTAL UNITS BUILT   23,421

TRACTOR SERIAL NUMBER DATA

Serial Numbers;	Year built;	Total units
224704 to 225109	Built in 1964	406
225110 to 227208	Built in 1965	2,099
227209 to 229224	Built in 1966	2,016
229225 to 231004	Built in 1967	1,780
231005 to 232980	Built in 1968	1,976
232981 to 234867	Built in 1969	1,887
234868 to 236826	Built in 1970	1,959
236827 to 238505	Built in 1971	1,679
238506 to 240580	Built in 1972	2,075
240581 to 242745	Built in 1973	2,165
242746 to 245650	Built in 1974	2,905
245651 to 248124	Built in 1975	2,474

TRACTOR MODEL "New" International Cub Tractor
PLACE OF MANUFACTURE Louisville

RATED HORSEPOWER	BELT	10	PTO
NEBRASKA TEST #	575		
ENGINE	C 60		
BORE & STROKE	2-5/8" x 2-3/4"		
AVERAGE SHIPPING WEIGHT			1,400lbs
TOTAL UNITS BUILT			5,561

TRACTOR SERIAL NUMBER DATA

Serial Numbers;	Year built;	Total units
248125 to 248617	Built in 1975	493
248618 to 250831	Built in 1976	2,214
250832 to 252108	Built in 1977	1,277
252109 to 253135	Built in 1978	1,027
253136 to 253685	Built in 1979	550

TRACTOR MODEL   Cub Lo-Boy
PLACE OF MANUFACTURE Louisville
RATED HORSEPOWER   BELT   10   PTO
NEBRASKA TEST #   575
ENGINE   C60
BORE & STROKE   2-5/8" x 2-3/4"
AVERAGE SHIPPING WEIGHT   1,650 lbs
TOTAL UNITS BUILT   25,498

TRACTOR SERIAL NUMBER DATA

Serial Numbers;	Year built;	Total units
501 to 2254	Built in 1955	1,754
2555 to 3928	Built in 1956	1,374
3929 to 6581	Built in 1957	2,652
6582 to 10566	Built in 1958	3,985
10567 to 12370	Built in 1959	1,804
12371 to 13909	Built in 1960	1,539
13910 to 15505	Built in 1961	1,596
15506 to 16439	Built in 1962	934
16440 to 17927	Built in 1963	1,488
17928 to 19405	Built in 1964	1,478
19406 to 21175	Built in 1965	1,770
21176 to 23114	Built in 1966	1,939
23115 to 24480	Built in 1967	1,366
24481 to 25998	Built in 1968	1,518

TRACTOR MODEL International Cub 154 Lo-Boy
PLACE OF MANUFACTURE Louisville
RATED HORSEPOWER PTO 18
NEBRASKA TEST # No Test
ENGINE C60
BORE & STROKE 2-5/8" x 2-3/4"
AVERAGE SHIPPING WEIGHT 1,500 lbs
TOTAL UNITS BUILT 29,172

TRACTOR SERIAL NUMBER DATA

Serial Numbers;	Year built;	Total units
7505 to 8272	Built in 1968	768
8273 to 15501	Built in 1969	7,229
15502 to 20331	Built in1970	4,829
20332 to 23342	Built in 1971	3,011
23343 to 27537	Built in 1972	4,195
27538 to 31765	Built in 1973	4,228
31766 to 36676	Built in 1974	4,911

TRACTOR MODEL	International Cub 184 Lo-Boy
PLACE OF MANUFACTURE	Louisville
RATED HORSEPOWER	PTO 20
NEBRASKA TEST #	No Test
ENGINE	C60
BORE & STROKE	2-5/8" x 2-3/4"
AVERAGE SHIPPING WEIGHT	1,550 lbs
TOTAL UNITS BUILT	8,537

TRACTOR SERIAL NUMBER DATA

Serial Numbers;	*Year built;*	*Total units*
43802 to 46162	Built in 1977	2,360
46163 to 48029	Built in 1978	1,866
48030 to 52339	Built in 1979	4,309

TRACTOR MODEL International Cub 185 Lo-Boy
PLACE OF MANUFACTURE Louisville
RATED HORSEPOWER PTO 18
NEBRASKA TEST # No Test
ENGINE C60
BORE & STROKE 2-5/8" x 2-3/4"
AVERAGE SHIPPING WEIGHT 1,500 lbs
TOTAL UNITS BUILT 6,347

TRACTOR SERIAL NUMBER DATA

Serial Numbers;	Year built;	Total units
37001 to 37315	Built in 1974	315
37316 to 41240	Built in 1975	3,925
41241 to 43347	Built in 1976	2,107

TRACTOR MODEL   A Farmall

PLACE OF MANUFACTURE

      Chicago Tractor Works & Louisville, Kentucky

RATED HORSEPOWER   BELT   18

NEBRASKA TEST # 329 Gasoline   #330 Distillate

ENGINE   C 113

BORE & STROKE   3x4

AVERAGE SHIPPING WEIGHT   2,200lbs

TOTAL UNITS BUILT 220,329 Chicago;   8,736 Louisville

TRACTOR SERIAL NUMBER DATA

Serial Numbers;	Year built;	Total units
501 to 6743	Built in 1939	6,243
6744 to 41499	Built in 1940	34,756
41500 to 80738	Built in 1941	39,239
80739 to 96389	Built in 1942	15,651
No Production	Built in 1943	NONE
96390 to 113217	Built in 1944	16,828
113218 to 146699	Built in 1945	33,482
146700 to 182963	Built in1946	36,264
182964 to 198298	Built in 1947	15,335
200001 to 220829	Built in 1947	20,829

TRACTOR MODEL   AV Farmall Hi-Clearance
PLACE OF MANUFACTURE
        Chicago Tractor Works & Louisville, Kentucky
RATED HORSEPOWER   BELT   18
NEBRASKA TEST # 329 Gasoline   #330 Distillate
ENGINE   C 113
BORE & STROKE   3x4
AVERAGE SHIPPING WEIGHT   3,000 lbs
TOTAL UNITS BUILT   506 at Louisville

TRACTOR SERIAL NUMBER DATA
Serial Numbers;   *Year built;*   *Total units*

48562 First AV Produced Built in 1941

All other serial number ranges are same as Farmall A

TRACTOR MODEL   International AI
PLACE OF MANUFACTURE
       Tractor Works Chicago & Louisville, Kentucky
RATED HORSEPOWER   BELT   18
NEBRASKA TEST #   No Test
ENGINE   C 113
BORE & STROKE   3x4
AVERAGE SHIPPING WEIGHT   3,500 lbs
TOTAL UNITS BUILT   838 at Louisville

TRACTOR SERIAL NUMBER DATA
Serial Numbers;   *Year built;*   *Total units*

57284 First AI Produced   Built in 1941

Other Serial Number Ranges are same as Farmall A

TRACTOR MODEL   Super A Farmall
PLACE OF MANUFACTURE Louisville, Kentucky
RATED HORSEPOWER BELT 20
NEBRASKA TEST #   None
ENGINE   C 113 & C123
BORE & STROKE   3x4 3-1/8 x 4
AVERAGE SHIPPING WEIGHT   2,400 Lbs
TOTAL UNITS BUILT   105,679

TRACTOR SERIAL NUMBER DATA

Serial Numbers;	Year built;	Total units
250001 to 250081	Built in 1947	81
250082 to 268195	Built in 1948	18,114
268196 to 281268	Built in 1949	13,073
281269 to 300125	Built in 1950	18,857
300126 to 324469	Built in 1951	24,344
324470 to 336879	Built in 1952	12,410
336880 to 353347	Built in 1953	16,468
353348 to 355679	Built in 1954	2,332

TRACTOR MODEL   Super AV Farmall (Hi-Clearance)
PLACE OF MANUFACTURE Louisville, Kentucky
RATED HORSEPOWER   **BELT**   20
NEBRASKA TEST #   None
ENGINE   C113   C123
BORE & STROKE   3 x 4   3-1/8 x 4
AVERAGE SHIPPING WEIGHT   2,600lbs
TOTAL UNITS BUILT   6,559

TRACTOR SERIAL NUMBER DATA
Serial Numbers;   *Year built;*   *Total units*

See Super A Farmall Serial Number List

TRACTOR MODEL   Super AI International
PLACE OF MANUFACTURE Louisville, KY
RATED HORSEPOWER   BELT   20
NEBRASKA TEST #   None
ENGINE   C113   C123
BORE & STROKE   3 x 4   3-1/8 x 4
AVERAGE SHIPPING WEIGHT   2,600 lbs
TOTAL UNITS BUILT   5,119

TRACTOR SERIAL NUMBER DATA

Serial Numbers;	*Year built;*	*Total units*

See Super A Farmall Serial Number List

TRACTOR MODEL   Super A-1
PLACE OF MANUFACTURE Louisville, KY
RATED HORSEPOWER   BELT   20
NEBRASKA TEST #   None
ENGINE   C113   C123
BORE & STROKE   3 x 4   3-1/8 x 4
AVERAGE SHIPPING WEIGHT   2,400 lbs

TOTAL UNITS BUILT

1,672	as Super A-1
58	as Super AI-1
228	as Super AV-1

TRACTOR SERIAL NUMBER DATA

Serial Numbers;	*Year built;*	*Total units*
356001 to 357958	Built in 1954	1,958

TRACTOR MODEL   Farmall B
PLACE OF MANUFACTURE
          Tractor Works, Chicago & Louisville, KY
RATED HORSEPOWER   BELT   18
NEBRASKA TEST # 331 Gasoline   #332 Distillate
ENGINE   C 113
BORE & STROKE   3 x 4
AVERAGE SHIPPING WEIGHT   3,700 lbs
TOTAL UNITS BUILT   8,781 at Louisville

TRACTOR SERIAL NUMBER DATA
Serial Numbers;   *Year built;*   *Total units*
3675   First Farmall B built in 1939

All other Serial Number ranges are same as Farmall A

TRACTOR MODEL   Farmall BN (Narrow)
PLACE OF MANUFACTURE
          Tractor Works Chicago & Louisville, KY
RATED HORSEPOWER   BELT   18
NEBRASKA TEST # 331 Gasoline   #332 Distillate
ENGINE   C 113
BORE & STROKE   3 x 4
AVERAGE SHIPPING WEIGHT   3,600 lbs
TOTAL UNITS BUILT   1,968 at Louisville

TRACTOR SERIAL NUMBER DATA

Serial Numbers;	Year built;	Total units
39715-X7 to 48349-X7	Built as BN	8,635

48459   Next production BN starting number

Serial Number Ranges are same as Farmall A

TRACTOR MODEL   Farmall Super B
PLACE OF MANUFACTURE Chicago Tractor Works
RATED HORSEPOWER
NEBRASKA TEST #
ENGINE
BORE & STROKE
AVERAGE SHIPPING WEIGHT
TOTAL UNITS BUILT

TRACTOR SERIAL NUMBER DATA
Serial Numbers;   *Year built;*   *Total units*

EXPERIMENTAL Tractor
Cancelled Prior to Production

TRACTOR MODEL   Farmall C
PLACE OF MANUFACTURE Louisville, KY
RATED HORSEPOWER   BELT   21
NEBRASKA TEST #   395
ENGINE   C 113   C123
BORE & STROKE   3 x 4   3-1/8 x 4
AVERAGE SHIPPING WEIGHT   2,900 lbs
TOTAL UNITS BUILT   79,932

TRACTOR SERIAL NUMBER DATA

Serial Numbers;	Year built;	Total units
501 to 22623	Built in 1948	22,123
22624 to 47009	Built in 1949	24,386
47010 to 71879	Built in 1950	24,870
71880 to 80432	Built in 1951	8,553

TRACTOR MODEL   Farmall Super C
PLACE OF MANUFACTURE Louisville, KY
RATED HORSEPOWER   **BELT**   23
NEBRASKA TEST #   458
ENGINE   C123
BORE & STROKE   3-1/8 x 4
AVERAGE SHIPPING WEIGHT   2,900 lbs
TOTAL UNITS BUILT   98,310

TRACTOR SERIAL NUMBER DATA

Serial Numbers;	*Year built;*	*Total units*
100001 to 131156	Built in 1951	31,156
131157 to 159129	Built in 1952	27,973
159130 to 187787	Built in 1953	28,658
187788 to 198310	Built in 1954	10,523

TRACTOR MODEL Farmall H & Farmall HV (Hi-Clear)
PLACE OF MANUFACTURE Farmall
RATED HORSEPOWER   BELT   26
NEBRASKA TEST # 333 Gasoline   334 Distillate
ENGINE   C 152
BORE & STROKE 3-3/8 x 4-1/4
AVERAGE SHIPPING WEIGHT   5,500 lbs
TOTAL UNITS BUILT   391,230

TRACTOR SERIAL NUMBER DATA

Serial Numbers;	Year built;	Total units
501 to 10652	Built in 1939	10,152
10653 to 52386	Built in 1940	41,734
52387 to 93236	Built in 1941	40,850
93237 to 122589	Built in 1942	29,353
122590 to 150250	Built in 1943	27,661
150251 to 186122	Built in 1944	35,872
186123 to 214819	Built in 1945	28,697
214820 to 241142	Built in 1946	26,323
241143 to 268990	Built in 1947	27,848
268991 to 300875	Built in 1948	31,885
300876 to 327974	Built in 1949	27,099
327975 to 351922	Built in 1950	23,948
351923 to 375860	Built in 1951	23,938
375861 to 390499	Built in 1952	14,639
390500 to 391730	Built in 1953	1,231

First Farmall H   built July-3-1939
First HV Built   SN 93237

TRACTOR MODEL   Farmall Super H and Super HV
PLACE OF MANUFACTURE Farmall
RATED HORSEPOWER   BELT   33
NEBRASKA TEST #   492
ENGINE   C 164
BORE & STROKE   3.5 x 4.25
AVERAGE SHIPPING WEIGHT   4,300 lbs
TOTAL UNITS BUILT   28,785

TRACTOR SERIAL NUMBER DATA

Serial Numbers;	*Year built;*	*Total units*
501 to 22201	Built in 1953	21,701
22202 to 29285	Built in 1954	7,084

Last Super H	SN 29285
Last Super HV	SN 27567

TRACTOR MODEL Farmall M & Farmall MV (Hi-Clear)
PLACE OF MANUFACTURE Farmall
RATED HORSEPOWER BELT 36
NEBRASKA TEST # 327 Distillate #328 Gasoline
ENGINE C 248
BORE & STROKE 3-7/8 x 5.25
AVERAGE SHIPPING WEIGHT 6,700 lbs
TOTAL UNITS BUILT 297,718

TRACTOR SERIAL NUMBER DATA

Serial Numbers;	Year built;	Total units
501 to 7239	Built in 1939	6,739
7240 to 25370	Built in 1940	18,131
25371 to 50987	Built in 1941	25,617
50988 to 60010	Built in 1942	9,023
60011 to 67423	Built in 1943	7,413
67424 to 88084	Built in 1944	20,661
88085 to 105563	Built in 1945	17,479
105564 to 122822	Built in1946	17,259
122823 to 151707	Built in 1947	28,885
151708 to 180513	Built in 1948	28,806
180514 to 213578	Built in 1949	33,065
213579 to 247517	Built in 1950	33,939
247518 to 290922	Built in 1951	43,405
290923 to 298218	Built in 1952	7,296

First LPG M tractor SN 287796

The 1 Millionth Farmall Series tractor made
was an M, built in 1947.

The 1 Millionth tractor made at the Farmall Plant
was an M built in April 1951 SN 262579

First Farmall M built July-7-1939

TRACTOR MODEL Farmall MD & Farmall MDV

(Diesel Hi- Clearance)

PLACE OF MANUFACTURE Farmall

RATED HORSEPOWER   BELT   35

NEBRASKA TEST #   368

ENGINE   D248

BORE & STROKE   3-7/8 x 5.25

AVERAGE SHIPPING WEIGHT   6,900 lbs

TOTAL UNITS BUILT

TRACTOR SERIAL NUMBER DATA

Serial Numbers;   *Year built;*   *Total units*

See Farmall M and MV list

25371	First MD Built
88085	First MDV Built
298218	Last MD Built
285385	Last MDV Built

TRACTOR MODEL Farmall Super M & Farmall Super MV
(Hi-Clear)

PLACE OF MANUFACTURE Farmall and Louisville, KY

RATED HORSEPOWER   BELT   47

NEBRASKA TEST #475 Gasoline   #484 LPG

ENGINE   C 264

BORE & STROKE   4 x 5.25

AVERAGE SHIPPING WEIGHT   5,600 lbs

TOTAL UNITS BUILT Farmall- 52,127; Louisville- 12,541

TRACTOR SERIAL NUMBER DATA

Serial Numbers;	Year built;	Total units
Farmall Built Tractors		
501 to 12515	Built in 1952	12,015
12516 to 51976	Built in 1953	39,461
51977 to 52627	Built in 1954	651
Louisville Built Tractors		
500001 to 501905	Built in 1952	1,905
501906 to 512541	Built in 1953	10,635

TRACTOR MODEL   Farmall Super MD
                               Farmall Super MDV (Hi-Clearance)
PLACE OF MANUFACTURE Farmall and Louisville, KY
RATED HORSEPOWER   BELT   47
NEBRASKA TEST #   477
ENGINE   D 264
BORE & STROKE   4 x 5.25
AVERAGE SHIPPING WEIGHT   5,800 lbs
TOTAL UNITS BUILT

TRACTOR SERIAL NUMBER DATA
Serial Numbers;   *Year built;*   *Total units*

See Super M and Super MV serial number List

TRACTOR MODEL   Super MTA and Super MV-TA
PLACE OF MANUFACTURE Farmall
RATED HORSEPOWER   BELT   47
NEBRASKA TEST #   No Test
ENGINE   C 264
BORE & STROKE   4 x 5.25
AVERAGE SHIPPING WEIGHT   5,600lbs
TOTAL UNITS BUILT   23,521

TRACTOR SERIAL NUMBER DATA

Serial Numbers;	*Year built;*	*Total units*
60001 to 83522	Built in 1954	23,521
61726	First Super MV-TA	
82776	Last Super M-TAV	
83522	Last Super MTA	

TRACTOR MODEL   Super MTAD and Super MV-TAD
PLACE OF MANUFACTURE Farmall
RATED HORSEPOWER   BELT   47
NEBRASKA TEST #   No Test
ENGINE   D 264
BORE & STROKE   4 x 5.25
AVERAGE SHIPPING WEIGHT   5,700lbs
TOTAL UNITS BUILT   21,848

TRACTOR SERIAL NUMBER DATA

Serial Numbers;	Year built;	Total units
60004 to 81848	Built in 1954	21,848

TRACTOR MODEL   W4 Standard Tractor
PLACE OF MANUFACTURE Farmall
RATED HORSEPOWER   BELT   34
NEBRASKA TEST # 342 Gasoline   #354 Distillate
ENGINE   C 152
BORE & STROKE   3-3/8 x 4.25
AVERAGE SHIPPING WEIGHT   5,600 lbs
TOTAL UNITS BUILT   33,676

TRACTOR SERIAL NUMBER DATA

Serial Numbers;	Year built;	Total units
501 to 942	Built in 1940	442
943 to 4055	Built in 1941	3,113
4056 to 5692	Built in 1942	1,637
5693 to 7592	Built in 1943	1,900
7593 to 11170	Built in 1944	3,578
11171 to 13933	Built in 1945	2,763
13934 to 16021	Built in 1946	2,088
16022 to 18879	Built in 1947	2,858
18880 to 21911	Built in 1948	3,032
21912 to 24469	Built in 1949	2,558
24470 to 28166	Built in 1950	3,697
28167 to 31213	Built in 1951	3,047
31214 to 33066	Built in 1952	1,853
33067 to 34176	Built in 1953	1,110

TRACTOR MODEL   O4 and OS4 Standard Tractor
PLACE OF MANUFACTURE Farmall
RATED HORSEPOWER   BELT   34
NEBRASKA TEST #   See W4
ENGINE   C152
BORE & STROKE   3-3/8 x 4.25
AVERAGE SHIPPING WEIGHT   5,600lbs
TOTAL UNITS BUILT

TRACTOR SERIAL NUMBER DATA

Serial Numbers;	*Year built;*	*Total units*
See W4 Serial Number Listing		

696	First O4 serial	
34043	First OS4 serial	
33753	Final OS4 serial	

TRACTOR MODEL Super W-4
PLACE OF MANUFACTURE Farmall
RATED HORSEPOWER BELT 34
NEBRASKA TEST # 491
ENGINE C 164
BORE & STROKE 3.5 x 4.25
AVERAGE SHIPPING WEIGHT 4,200lbs
TOTAL UNITS BUILT 2,792

TRACTOR SERIAL NUMBER DATA

Serial Numbers;	Year built;	Total units
501 to 2667	Built in 1953	2,167
2668 to 3292	Built in 1954	625

TRACTOR MODEL W-6
PLACE OF MANUFACTURE Farmall
RATED HORSEPOWER BELT 36
NEBRASKA TEST # 335
ENGINE C 248
BORE & STROKE 3-7/8 x 5.25
AVERAGE SHIPPING WEIGHT 5,500lbs
TOTAL UNITS BUILT 45,511

TRACTOR SERIAL NUMBER DATA

Serial Numbers;	Year built;	Total units
501 to 1224	Built in 1940	724
1225 to 3717	Built in 1941	2,493
3718 to 5056	Built in 1942	1,339
5057 to 6312	Built in 1943	1,256
6313 to 9495	Built in 1944	3,183
9496 to 14152	Built in 1945	4,657
14153 to 17791	Built in 1946	3,639
17792 to 22980	Built in 1947	5,188
22981 to 28703	Built in 1948	5,723
28704 to 33697	Built in 1949	4,994
33698 to 38517	Built in 1950	4,820
38518 to 44317	Built in 1951	5,800
44318 to 45273	Built in 1952	956
45274 to 46279	Built in 1953	1,006

TRACTOR MODEL WD-6
PLACE OF MANUFACTURE Farmall
RATED HORSEPOWER BELT 36
NEBRASKA TEST # 356
ENGINE D 248 Diesel
BORE & STROKE 3-7/8 x 5.25
AVERAGE SHIPPING WEIGHT 5,700lbs
TOTAL UNITS BUILT

TRACTOR SERIAL NUMBER DATA
Serial Numbers; *Year built;* *Total units*
See W-6 Serial Number Listing

45416 Last WD-6 Built

TRACTOR MODEL   Super W-6 & Super W-6 Diesel
PLACE OF MANUFACTURE Farmall
RATED HORSEPOWER   BELT   46
NEBRASKA TEST # 476 Gasoline   478 Diesel 485
ENGINE   C 264 Gasoline   D 264 Diesel
BORE & STROKE   4 x 5.25
AVERAGE SHIPPING WEIGHT   5,500lbs
TOTAL UNITS BUILT   8,584

TRACTOR SERIAL NUMBER DATA

Serial Numbers;	Year built;	Total units
501 to 2907	Built in 1952	2,407
2908 to 8996	Built in 1953	6,089
8997 to 9084	Built in 1954	88

TRACTOR MODEL   OS-6 Orchard Special /
                             ODS-6 Orchard Special Diesel

PLACE OF MANUFACTURE Farmall

RATED HORSEPOWER   BELT   36

NEBRASKA TEST #   No Test

ENGINE   C 248 Gasoline   D 248 Diesel

BORE & STROKE   3-7/8 x 5.25

AVERAGE SHIPPING WEIGHT   5,700lbs

TOTAL UNITS BUILT

TRACTOR SERIAL NUMBER DATA

Serial Numbers;   *Year built;*   *Total units*

      See W-6 Listing

811   First O-6 Built
45782   Last O-6 Built

9062   First OS-6 Built
45843   Last OS-6 Built

45281   Last OSD-6 Built

TRACTOR MODEL   Super W-6TA & Super W-6TA Diesel
PLACE OF MANUFACTURE Farmall
RATED HORSEPOWER   BELT   46
NEBRASKA TEST #   No Test
ENGINE   C 264 Gasoline   D 264 Diesel
BORE & STROKE   4 x 5.25
AVERAGE SHIPPING WEIGHT   5,500lbs
TOTAL UNITS BUILT   12,006

TRACTOR SERIAL NUMBER DATA
Serial Numbers;   *Year built;*   *Total units*
1001 to 13005   Built in 1954   12,005

13005   Last Super W6-TA

TRACTOR MODEL W-9 / WD-9 Diesel /
           WR-9 Rice Special / WDR-9
PLACE OF MANUFACTURE Milwaukee Works and Farmall
RATED HORSEPOWER   BELT   47
NEBRASKA TEST #s

369 Gasoline   370 Diesel   371 Distillate
ENGINE   C 335 Gas   D 335 Diesel
BORE & STROKE   4.4 x 5.5
AVERAGE SHIPPING WEIGHT   6,300lbs
TOTAL UNITS BUILT   67,419

TRACTOR SERIAL NUMBER DATA

Serial Numbers;	Year built;	Total units
501 to 577	Built in 1940	77
578 to 2992	Built in 1941	2,415
2993 to 3650	Built in 1942	658
3651 to 5393	Built in 1943	1,743
5394 to 11458	Built in 1944	6,065
11459 to 17288	Built in 1945	5,830
17289 to 22713	Built in 1946	5,425
22714 to 29206	Built in 1947	6,493
29207 to 36158	Built in 1948	6,952
36159 to 45550	Built in 1949	9,392
45551 to 51738	Built in 1950	6,188
51739 to 59406	Built in 1951	7,668
59407 to 64013	Built in 1952	4,607
64014 to 67919	Built in 1953	3,905

66870   Final W-9 Built
67898   Final WD-9 Built
67506   Final WDR-9 Built
67919   Final WR-9 Built

TRACTOR MODEL   WR-9S Rice Special
PLACE OF MANUFACTURE Farmall
RATED HORSEPOWER   BELT   50
NEBRASKA TEST #   No Test
ENGINE   C 335 Gasoline
BORE & STROKE   4.4 x 5.5
AVERAGE SHIPPING WEIGHT   6,300lbs
TOTAL UNITS BUILT   279

TRACTOR SERIAL NUMBER DATA

Serial Numbers;	Year built;	Total units
501 to 549	Built in 1953	49
550 to 721	Built in 1954	172
722 to 754	Built in 1955	33
755 to 779	Built in 1956	25

TRACTOR MODEL Super WD-9 /
                   Super WDR-9 Rice Special
PLACE OF MANUFACTURE Farmall
RATED HORSEPOWER   BELT   65
NEBRASKA TEST #   518 Diesel
ENGINE   D 350
BORE & STROKE   4.5 x 5.5
AVERAGE SHIPPING WEIGHT   6,700lbs
TOTAL UNITS BUILT   6,753 SWD-9 / 6,749 SWDR-9

TRACTOR SERIAL NUMBER DATA

Serial Numbers;	Year built;	Total units
SWD-9		
501 to 1932	Built in 1953	1,432
1933 to 5229	Built in 1954	3,297
5230 to 6863	Built in 1955	1,634
6864 to 7253	Built in 1956	390
SWDR-9		
501 to 1934	Built in 1953	1,434
1933 to 5237	Built in 1954	3,303
5238 to 6865	Built in 1955	1,628
6866 to 7249	Built in 1956	384

TRACTOR MODEL   100 Farmall
PLACE OF MANUFACTURE Louisville, KY
RATED HORSEPOWER   BELT   20
NEBRASKA TEST #   537
ENGINE   C 123
BORE & STROKE   3-1/8 x 4
AVERAGE SHIPPING WEIGHT   2,600lbs
TOTAL UNITS BUILT   18,440

TRACTOR SERIAL NUMBER DATA

Serial Numbers;	Year built;	Total units
501 to 1719	Built in 1954	1,219
1720 to 12894	Built in 1955	11,175
12895 to 18940	Built in 1956	6,046

TRACTOR MODEL   100 International
PLACE OF MANUFACTURE Louisville, KY
RATED HORSEPOWER   BELT   20
NEBRASKA TEST #   537
ENGINE   C 123
BORE & STROKE   3-1/8 x 4
AVERAGE SHIPPING WEIGHT   2,700lbs
TOTAL UNITS BUILT   135

TRACTOR SERIAL NUMBER DATA

Serial Numbers;	*Year built;*	*Total units*
501 to 503	Built in 1954	3
504 to 574	Built in 1955	71
575 to 635	Built in 1956	61

TRACTOR MODEL   130 Farmall & International
PLACE OF MANUFACTURE Louisville, KY
RATED HORSEPOWER   BELT   22
NEBRASKA TEST#   617
ENGINE SIZE   C 123
BORE & STROKE   3-1/8 X 4
AVERAGE SHIPPING WEIGHT   2,800lbs
TOTAL UNITS BUILT   9,709

TRACTOR SERIAL NUMBER DATA

Serial Numbers;	Year built;	Total units
501 to 1119	Built in 1956	619
1120 to 8362	Built in 1957	7,243
8363 to 10209	Built in 1958	1,847

570 units were built as Hi-Clearance Tractors

TRACTOR MODEL	140 Farmall & International
PLACE OF MANUFACTURE	Louisville , KY
RATED HORSEPOWER	BELT 23
NEBRASKA TEST #	666
ENGINE SIZE	C 123
BORE & STROKE	3-1/8x 4
AVERAGE SHIPPING WEIGHT	3,000 lbs.
TOTAL UNITS BUILT	66,290

TRACTOR SERIAL NUMBER DATA

Serial Numbers;	*Year built;*	*Total units*
501 to 2010	Built in 1958	1,510
2011 to 8081	Built in 1959	6,071
8082 to 11167	Built in 1960	3,086
11168 to 16636	Built in 1961	5,469
11637 to 21180	Built in 1962	4,544
21181 to 25386	Built in 1963	4,206
25387 to 28407	Built in 1964	3,021
28408 to 31284	Built in 1965	2,877
31285 to 34817	Built in 1966	3,533
34818 to 37351	Built in 1967	2,534
37352 to 39905	Built in 1968	2,554
39906 to 42299	Built in 1969	2,393
42300 to 44423	Built in 1970	2,124
44424 to 44604	Built in 1971	181
46605 to 48506	Built in 1972	3,902
48507 to 50719	Built in 1973	2,213
50720 to 54272	Built in 1974	3,553
54273 to 57295	Built in 1975	3,023
57296 to 60838	Built in 1976	3,543
60839 to 63110	Built in 1977	2,272
63111 to 64543	Built in 1978	1,433
64544 to 66790	Built in 1979*	2,247

2,861 units were built as Hi- Clearance
5,865 units were built as Internationals

TRACTOR MODEL 200 Farmall
PLACE OF MANUFACTURE Louisville, KY
RATED HORSEPOWER BELT 24
NEBRASKA TEST # 536
ENGINE SIZE C 123
BORE & STROKE 3-1/8 x 4
AVERAGE SHIPPING WEIGHT 3,100 lbs.
TOTAL UNITS BUILT 15,198

TRACTOR SERIAL NUMBER DATA

Serial Numbers;	Year built;	Total units
501 to 1031	Built in 1954	531
1032 to 10903	Built in 1955	9,872
10904 to 15698	Built in 1956	4,795

TRACTOR MODEL 230 Farmall
PLACE OF MANUFACTURE Louisville, KY
RATED HORSEPOWER BELT 28
NEBRASKA TEST# 616
ENGINE SIZE C 123
BORE & STROKE 3-1/8 X 4
AVERAGE SHIPPING WEIGHT 3,200 lbs.
TOTAL UNITS BUILT 7,171

TRACTOR SERIAL NUMBER DATA

Serial Numbers;	Year built;	Total units
501 to 814	Built in 1956	314
815 to 6826	Built in 1957	6,326
6827 to 7671	Built in 1958	845

TRACTOR MODEL 240 Farmall
PLACE OF MANUFACTURE Louisville, KY
RATED HORSEPOWER BELT 30
NEBRASKA TEST# 667 row crop
ENGINE SIZE C 123 Gasoline
BORE & STROKE 3-1/8 X 4
AVERAGE SHIPPING WEIGHT 3,300 lbs.
TOTAL UNITS BUILT 3,624

TRACTOR SERIAL NUMBER DATA

Serial Numbers;	Year built;	Total units
501 to 1776	Built in 1958	1,276
1777 to 3414	Built in 1959	1,648
3415 to 3988	Built in 1960	574
3989 to 4124	Built in 1961	136

TRACTOR MODEL 240 International

PLACE OF MANUFACTURE Louisville, KY

RATED HORSEPOWER BELT 30

NEBRASKA TEST # 668 utility

ENGINE SIZE C 123 Gasoline

BORE & STROKE 3-1/8 X 4

AVERAGE SHIPPING WEIGHT 3,300 lbs.

TOTAL UNITS BUILT 10,289

TRACTOR SERIAL NUMBER DATA

Serial Numbers;	Year built;	Total units
501 to 4834	Built in 1958	4,334
4835 to 8627	Built in 1959	3,793
8628 to 10078	Built in 1960	1,451
10079 to 10726	Built in 1961	648
10727 to 10789	Built in 1962	63

TRACTOR MODEL   300 Farmall
PLACE OF MANUFACTURE Farmall
RATED HORSEPOWER   BELT   38
NEBRASKA TEST#   538 Gasoline   573 LPG
ENGINE SIZE   C 169 Gasoline & LPG
BORE & STROKE   3-9/16 X 4.25"
AVERAGE SHIPPING WEIGHT   4,000 lbs.
TOTAL UNITS BUILT   29,078

TRACTOR SERIAL NUMBER DATA

Serial Numbers;	*Year built;*	*Total units*
501 to 1778	Built in 1954	1,278
1779 to 23223	Built in 1955	21,445
23224 to 29578	Built in 1956	6,355

TRACTOR MODEL 300 International
PLACE OF MANUFACTURE Farmall
RATED HORSEPOWER BELT 38
NEBRASKA TEST # 539
ENGINE SIZE C 169 Gasoline & LPG
BORE & STROKE 3-9/16 X 4.25"
AVERAGE SHIPPING WEIGHT 4,800lbs.
TOTAL UNITS BUILT 30,852

TRACTOR SERIAL NUMBER DATA

Serial Numbers;	Year built;	Total units
501 to 20218	Built in 1955	19,718
20219 to 31352	Built in 1956	11,134

I-300 Tractor SN 501 is IH's 3 Millionth Tractor
built April 1955

TRACTOR MODEL   330 International
PLACE OF MANUFACTURE Farmall
RATED HORSEPOWER   BELT   34
NEBRASKA TEST #   634
ENGINE SIZE   C 135 Gasoline
BORE & STROKE   3.25 X 4-1/16"
AVERAGE SHIPPING WEIGHT   4,000 lbs.
TOTAL UNITS BUILT   4,263

TRACTOR SERIAL NUMBER DATA

Serial Numbers;	*Year built;*	*Total units*
501 to 1487	Built in 1957	987
1488 to 4763	Built in 1958	3,276

TRACTOR MODEL   340 Farmall
PLACE OF MANUFACTURE Farmall
RATED HORSEPOWER   BELT   39
NEBRASKA TEST#   665 Gasoline   775 Diesel
ENGINE SIZE   C135 Gasoline   D166 Diesel
BORE & STROKE 3.25 X 4-1/16"   3-11/16 X 3-7/8"
AVERAGE SHIPPING WEIGHT   4,500 lbs.
TOTAL UNITS BUILT   7,211

TRACTOR SERIAL NUMBER DATA

Serial Numbers;	Year built;	Total units
501 to 2722	Built in 1958	2,222
2723 to 5410	Built in 1959	2,688
5411 to 6641	Built in 1960	1,231
6642 to 7625	Built in 1961	984
7626 to 7698	Built in 1962	73
7699 to 7711	Built in 1963	13

TRACTOR MODEL 340 International / 340 Utility
PLACE OF MANUFACTURE Farmall
RATED HORSEPOWER BELT 39
NEBRASKA TEST # 665 Gasoline 775 Diesel
ENGINE SIZE C 135 Gasoline D 166 Diesel
BORE & STROKE 3.25 X 4-1/16" 3-11/16 X 3-7/8"
AVERAGE SHIPPING WEIGHT 4,300 lbs
TOTAL UNITS BUILT 11,738

TRACTOR SERIAL NUMBER DATA

Serial Numbers;	Year built;	Total units
501 to 2466	Built in 1958	1,966
2467 to 5740	Built in 1959	3,274
5741 to 8735	Built in 1960	2,995
8736 to 11140	Built in 1961	2,405
11141 to 12031	Built in 1962	891
12032 to 12238	Built in 1963	207

TRACTOR MODEL   350 Farmall
PLACE OF MANUFACTURE Farmall
RATED HORSEPOWER   BELT   40
NEBRASKA TEST #611 Gasoline   622 LPG 609 Diesel
ENGINE SIZE C175 Gasoline & LPG D193 Diesel
BORE & STROKE 3-5/8 x 4.25   3.75 x 4.75
AVERAGE SHIPPING WEIGHT   4,500lbs
TOTAL UNITS BUILT   16,715

TRACTOR SERIAL NUMBER DATA

Serial Numbers;	*Year built;*	*Total units*
501 to 1003	Built in 1956	503
1004 to 14174	Built in 1957	13,171
14175 to 17215	Built in 1958	3,041

501	First F-350 Gas
581	First F-350 Diesel
535	First F-350 LPG
533	First F-350 HiClear

TRACTOR MODEL   350 International
PLACE OF MANUFACTURE Farmall
RATED HORSEPOWER   BELT   43
NEBRASKA TEST # z   615 Gasoline   610 Diesel
ENGINE SIZE C 175 Gasoline & LPG   D193 Diesel
BORE & STROKE   3-5/8 x 4.25   3.75 x 4.75
AVERAGE SHIPPING WEIGHT   4,200lbs
TOTAL UNITS BUILT   17,846

TRACTOR SERIAL NUMBER DATA

Serial Numbers;	Year built;	Total units
501 to 1962	Built in 1958	1,462
1963 to 15048	Built in 1957	13,086
15049 to 18346	Built in 1958	3,298

501	First I-350 Gas
739	First I-350 Diesel
590	First I-350 LPG

TRACTOR MODEL   400 Farmall

PLACE OF MANUFACTURE Farmall

RATED HORSEPOWER   BELT   50

NEBRASKA TEST # 532 Gasoline   534 Diesel

ENGINE SIZE C 264 Gasoline & LPG D 264 Diesel

BORE & STROKE   4 x 5-1/4

AVERAGE SHIPPING WEIGHT   5,200lbs

TOTAL UNITS BUILT   38,361

TRACTOR SERIAL NUMBER DATA

Serial Numbers;	Year built;	Total units
501 to 2587	Built in 1954	2,087
2588 to 29064	Built in 1955	26,477
29065 to 38861	Built in 1956	9,797

TRACTOR MODEL   W400 (Wheatland)
PLACE OF MANUFACTURE Farmall
RATED HORSEPOWER   BELT   51
NEBRASKA TEST # 533 Gasoline #572 LPG   #535 Diesel
ENGINE SIZE C 264 Gasoline & LPG   D264 Diesel
BORE & STROKE   4 x 5-1/4
AVERAGE SHIPPING WEIGHT   5,500lbs
TOTAL UNITS BUILT   3,358

TRACTOR SERIAL NUMBER DATA

Serial Numbers;	Year built;	Total units
501 to 2186	Built in 1955	1,686
2187 to 3858	Built in 1956	1,672

TRACTOR MODEL   404 Farmall
PLACE OF MANUFACTURE Farmall & Louisville, KY
RATED HORSEPOWER   BELT   37
NEBRASKA TEST #   818
ENGINE SIZE   C 135 Gasoline & LPG
BORE & STROKE   3-1/4x 4-1/16
AVERAGE SHIPPING WEIGHT   3,556 lbs
TOTAL UNITS BUILT   2,575
                          2,240 units built at Louisville

TRACTOR SERIAL NUMBER DATA

Serial Numbers;	Year built;	Total units
501 to 825	Built in 1961	325
826 to 1935	Built in 1962	1,110
1936 to 2258	Built in 1963	323
2259 to 2567	Built in 1964	309
2568 to 2789	Built in 1965	222
2790 to 2979	Built in 1966	190
2980 to 3075	Built in 1967	96

TRACTOR MODEL   404 International
PLACE OF MANUFACTURE Farmall & Louisville, KY
RATED HORSEPOWER   BELT   37
NEBRASKA TEST #
ENGINE SIZE   135 Gasoline & LPG
BORE & STROKE   3-1/4 x 4-1/16
AVERAGE SHIPPING WEIGHT   3,427lbs
TOTAL UNITS BUILT   10,577

8,256 Units built at Louisville

TRACTOR SERIAL NUMBER DATA

Serial Numbers;	Year built;	Total units
501 to 1044	Built in 1961	544
1045 to 4204	Built in 1962	3,160
4205 to 6451	Built in 1963	2,247
6452 to 8291	Built in 1964	1,840
8292 to 9527	Built in 1965	1,235
9528 to 10533	Built in 1966	1,006
10534 to 11031	Built in 1967	497
11032 to 11077	Built in 1968	46

TRACTOR MODEL   424
PLACE OF MANUFACTURE Louisville, KY
RATED HORSEPOWER   BELT   37
NEBRASKA TEST # 908 Gasoline   #911 Diesel
ENGINE SIZE   C 146 Gasoline   BD 154 Diesel
BORE & STROKE   3-3/8 x 4-1/16   3.5 x 4
AVERAGE SHIPPING WEIGHT   3,600lbs
TOTAL UNITS BUILT   16,631

TRACTOR SERIAL NUMBER DATA

Serial Numbers;	Year built;	Total units
501 to 1401	Built in 1964	901
1402 to 7840	Built in 1965	6,439
7841 to 13626	Built in 1966	5,786
13627 to 17131	Built in 1967	3,505

TRACTOR MODEL 444

PLACE OF MANUFACTURE Louisville, KY

RATED HORSEPOWER BELT 38

NEBRASKA TEST # 985

ENGINE SIZE C 153 Gasoline BD 154 Diesel

BORE & STROKE 3-3/8 x 4.25 3.5 x 4

AVERAGE SHIPPING WEIGHT 4,100lbs

TOTAL UNITS BUILT 12,664

TRACTOR SERIAL NUMBER DATA

Serial Numbers;	Year built;	Total units
501 to 1189	Built in 1967	689
1190 to 5719	Built in 1968	4,530
5720 to 9009	Built in 1969	3,290
9010 to 12356	Built in 1970	3,347
12357 to 13164	Built in 1971	808

TRACTOR MODEL   W 450 (Wheatland)
PLACE OF MANUFACTURE Farmall
RATED HORSEPOWER   BELT   55
NEBRASKA TEST #   No Test
ENGINE SIZE C 281 Gasoline & LPG   D281 Diesel
BORE & STROKE   4-1/8 x 5.25
AVERAGE SHIPPING WEIGHT   6,300lbs
TOTAL UNITS BUILT   1,795

TRACTOR SERIAL NUMBER DATA

Serial Numbers;	Year built;	Total units
501 to 567	Built in 1956	67
568 to 1660	Built in 1957	1,093
1661 to 2295	Built in 1958	635

501	First IW-450 Gas
529	First IW-450 Diesel
517	First IW-450 LPG

TRACTOR MODEL 450 Farmall
PLACE OF MANUFACTURE Farmall
RATED HORSEPOWER BELT 55
NEBRASKA TEST # 612 Gasoline 620 LPG 608 Diesel
ENGINE SIZE C281 D281
BORE AND STROKE 4-1/8x 5.25"
AVERAGE SHIPPING WEIGHT 6,000lbs
TOTAL UNITS BUILT 25,566

TRACTOR SERIAL NUMBER DATA

Serial Numbers;	*Year built;*	*Total units*
501 to 1733	Built in 1956	1,233
1734 to 21870	Built in 1957	20,137
21871 to 26066	Built in 1958	4,196

501	First F-450 Gas
585	First F-450 Diesel
579	First F-450 LPG
596	First F-450 HiClear
2607	First F-450 Diesel HiClear

TRACTOR MODEL 454
PLACE OF MANUFACTURE Louisville, KY
RATED HORSEPOWER BELT 40
NEBRASKA TEST # 985 & 1096 Gasoline #1097 Diesel
ENGINE SIZE C 157 & C175 gasoline D179 Diesel
BORE AND STROKE 3-3/8x 4.39 (C157)
3-7/8 x 5.060
3-9/16 x 4.39 (C175)
AVERAGE SHIPPING WEIGHT 4,000 lbs
TOTAL UNITS BUILT 7,812

TRACTOR SERIAL NUMBER DATA

Serial Numbers;	Year built;	Total units
501 to 507	Built in 1970	7
508 to 4907	Built in 1971	4,399
4908 to 8063	Built in 1972	3,156
8064 to 8313	Built in 1973	250

TRACTOR MODEL   460 Farmall
PLACE OF MANUFACTURE Farmall
RATED HORSEPOWER   BELT   50
NEBRASKA TEST #670 Gasoline   676 LPG   672 Diesel
ENGINE SIZE   C221 Gasoline & LPG   D236 Diesel
BORE AND STROKE   3-9/16 x 3-11/16
                         3-11/16 x 3-11/16"
AVERAGE SHIPPING WEIGHT   5,500lbs
TOTAL UNITS BUILT   32,528

TRACTOR SERIAL NUMBER DATA

Serial Numbers;	Year built;	Total units
501 to 4764	Built in 1958	4,264
4765 to 16901	Built in 1959	12,137
16902 to 22621	Built in 1960	5,720
22622 to 28028	Built in 1961	5,407
28029 to 31551	Built in 1962	3,523
31552 to 33028	Built in 1963	1,477

501	First F-460 Gas
663	First F-460 Diesel
3379	First F-460 LPG
1940	First F-460 HiClear

TRACTOR MODEL   460 International
PLACE OF MANUFACTURE Farmall
RATED HORSEPOWER   **BELT**   50
NEBRASKA TEST # 674 Gasoline 677 LPG   673 Diesel
ENGINE SIZE   C221 Gasoline & LPG   D236 Diesel
BORE AND STROKE   3-9/16 x 3-11/16"
                        3-11/16 x 3-11/16"
AVERAGE SHIPPING WEIGHT   5,300lbs
TOTAL UNITS BUILT   11,411

TRACTOR SERIAL NUMBER DATA

Serial Numbers;	Year built;	Total units
501 to 2710	Built in 1958	2,210
2711 to 6882	Built in 1959	4,172
6883 to 9419	Built in 1960	2,537
9420 to 11618	Built in 1961	2,199
11619 to 11897	Built in 1962	279
11898 to 11911	Built in 1963	14

501	First I-460 Gas
579	First I-460 Diesel
2116	First I-460 LPG

TRACTOR MODEL   464
PLACE OF MANUFACTURE Louisville KY
RATED HORSEPOWER   BELT   45
NEBRASKA TEST # 1126 Gasoline   1127 Diesel
ENGINE SIZE   C 175   D 179
BORE AND STROKE 3-9/16 x 4.39" 3-7/8 x 5.060"
AVERAGE SHIPPING WEIGHT   4,600lbs
TOTAL UNITS BUILT   10,475

TRACTOR SERIAL NUMBER DATA

Serial Numbers;	Year built;	Total units
100003 to 102195	Built in 1973	2,193
102196 to 104774	Built in 1974	2,579
104775 to 106878	Built in 1975	2,104
106879 to 108714	Built in 1976	1,836
108715 to 110478	Built in 1977	1,764
110479 to --	Built in 1978	

TRACTOR MODEL   504 Farmall
PLACE OF MANUFACTURE Farmall
RATED HORSEPOWER   BELT   45
NEBRASKA TEST # 819 Gas 820 LPG   816 diesel
ENGINE SIZEC153 Gas and LPG   D 188 Diesel
BORE AND STROKE   3-3/8 x 4.25"
                         3-11/16 x 4.39"
AVERAGE SHIPPING WEIGHT   5,000 lbs
TOTAL UNITS BUILT   15,675

TRACTOR SERIAL NUMBER DATA

Serial Numbers;	Year built;	Total units
501 to 809	Built in 1961	309
810 to 6999	Built in 1962	6,190
7000 to 7731	Built in 1963	732
7732 to 10695	Built in 1964	2,964
10696 to 13595	Built in 1965	2,900
13596 to 15112	Built in 1966	1,517
15113 to 16114	Built in 1967	1,002
16115 to 16175	Built in 1968	61

501	First F-504 Gas
583	First F-504 Diesel
765	First F-504 LPG

TRACTOR MODEL   504 International
PLACE OF MANUFACTURE Farmall
RATED HORSEPOWER   BELT   45
NEBRASKA TEST #819 Gas 820 LPG   816 diesel
ENGINE SIZE C153 Gas and LPG   D188 Diesel
BORE AND STROKE   3-3/8 x 4.25"
                         3-11/16 x 4.39"
AVERAGE SHIPPING WEIGHT   4,200 lbs
TOTAL UNITS BUILT   19,938

TRACTOR SERIAL NUMBER DATA

Serial Numbers;	Year built;	Total units
501 to 3375	Built in 1962	2,875
3376 to 6796	Built in 1963	3,394
6796 to 10995	Built in 1964	4,200
10996 to 14694	Built in 1965	3,699
14695 to 17991	Built in 1966	3,297
17992 to 20391	Built in 1967	2,400
20392 to 20438	Built in 1968	47

501	First I-504 Gas
508	First I-504 Diesel
511	First I-504 LPG

TRACTOR MODEL   544 Farmall
PLACE OF MANUFACTURE Farmall
RATED HORSEPOWER BELT   54
NEBRASKA TEST # 983 Gas   1007 Hydrostatic
                         984 Diesel   1029 Hydro Diesel
ENGINE SIZE   C200 Gas   D239 Diesel
BORE AND STROKE 3-3/16 x 4.39" 3-7/8 x 5.060"
AVERAGE SHIPPING WEIGHT   6,000 lbs
TOTAL UNITS BUILT   5,800

TRACTOR SERIAL NUMBER DATA

Serial Numbers;	Year built;	Total units
10250 to 12540	Built in 1968	2,291
12541 to 13584	Built in 1969	1,044
13585 to 14506	Built in 1970	922
14507 to 15261	Built in 1971	755
15262 to 15737	Built in 1972	476
15738 to 16049	Built in 1973	312

1970 Year built tractors COULD BE an IH Golden Demonstrator

TRACTOR MODEL   544 International
PLACE OF MANUFACTURE Farmall
RATED HORSEPOWER BELT   54
NEBRASKA TEST # 983 Gasoline   1007 Hydrostatic
                    984 Diesel   1029 Hydro Diesel
ENGINE SIZE   C 200 Gasoline   D239 Diesel
BORE & STROKE 3-3/16 x 4.39"   3-7/8 x 5.060"
AVERAGE SHIPPING WEIGHT   5,300lbs
TOTAL UNITS BUILT   7,179

TRACTOR SERIAL NUMBER DATA

Serial Numbers;	Year built;	Total units
10250 to 12698	Built in 1968	2,449
12699 to 14588	Built in 1969	1,890
14589 to 16017	Built in 1970	1,429
16018 to 16837	Built in 1971	820
16838 to 17340	Built in 1972	503
17341 to 17428	Built in 1973	88

1970 Year built tractors COULD BE an IH Golden
Demonstrator

TRACTOR MODEL   560 Farmall
PLACE OF MANUFACTURE Farmall
RATED HORSEPOWER   BELT   61
NEBRASKA TEST # 671 Gasoline 675 LPG   669 Diesel
ENGINE SIZE C263 Gasoline & LPG   D-282 Diesel
BORE & STROKE 3-9/16 x 4.39"   3-11/16 x 4.39"
AVERAGE SHIPPING WEIGHT   6,000lbs
TOTAL UNITS BUILT   65,532

TRACTOR SERIAL NUMBER DATA

Serial Numbers;	Year built;	Total units
501 to 7340	Built in 1958	6,840
7341 to 29613	Built in 1959	22,272
29614 to 36124	Built in 1960	6,511
36125 to 47797	Built in 1961	11,673
47798 to 60277	Built in 1962	12,480
60278 to 66032	Built in 1963	5,755

501	First F-560 Gas
725	First F-560 Diesel
4981	First F-560 LPG
1422	First F-560 HiClear
5952	First F-560 HiClear LPG

TRACTOR MODEL   560 International
PLACE OF MANUFACTURE Farmall
RATED HORSEPOWER   BELT   62
NEBRASKA TEST # 671 Gasoline 675 LPG   669 Diesel
ENGINE SIZE   C263 Gasoline & LPG   D282 Diesel
BORE & STROKE 3-9/16 x 4.39"   3-11/16 x 4.39"
AVERAGE SHIPPING WEIGHT   6,300lbs
TOTAL UNITS BUILT   5,549

TRACTOR SERIAL NUMBER DATA

Serial Numbers;	*Year built;*	*Total units*
501 to 1209	Built in 1958	706
1210 to 3102	Built in 1959	1,893
3103 to 4031	Built in 1960	929
4032 to 4943	Built in 1961	912
4944 to 5597	Built in 1962	654
5598 to 6049	Built in 1963	452

501	First I-560 Gas
525	First I-560 Diesel
1071	First I-560 LPG

TRACTOR MODEL   International 600
PLACE OF MANUFACTURE Farmall
RATED HORSEPOWER   BELT   63
NEBRASKA TEST #
ENGINE SIZE   C350 Gasoline   D350 Diesel
BORE & STROKE   4.5 x 5.5"
AVERAGE SHIPPING WEIGHT   6,700lbs
TOTAL UNITS BUILT   1,485

TRACTOR SERIAL NUMBER DATA

Serial Numbers;	*Year built;*	*Total units*
501 to 1985	Built in 1956	1,485

TRACTOR MODEL   International 606
PLACE OF MANUFACTURE Farmall
RATED HORSEPOWER   PTO   54
NEBRASKA TEST # 825 Gas   826 Diesel
ENGINE SIZE   C221 Gasoline   D236 Diesel
BORE & STROKE 3-9/16 x 3-11/16 3-11/16 x 3-11/16
AVERAGE SHIPPING WEIGHT   4,900 lbs
TOTAL UNITS BUILT   7,439

TRACTOR SERIAL NUMBER DATA

Serial Numbers;	Year built;	Total units
501 to 502	Built in 1961	2
503 to 1594	Built in 1962	1,092
1595 to 3214	Built in 1963	1,619
3215 to 4968	Built in 1964	1,753
5041 to 6849	Built in 1965	1,808
6960 to 7912	Built in 1966	952
7922 to 7939	Built in 1967	17

501	First I-606 Gas
502	First I-606 Diesel
922	First I-606 LPG

TRACTOR MODEL   International 650
PLACE OF MANUFACTURE Farmall
RATED HORSEPOWER   BELT   63
NEBRASKA TEST #   618 Gas   621 LPG
ENGINE SIZE C350 Gasoline & LPG D350 Diesel
BORE & STROKE   4.5 x 5.5"
AVERAGE SHIPPING WEIGHT   6,800lbs
TOTAL UNITS BUILT

TRACTOR SERIAL NUMBER DATA

Serial Numbers;	Year built;	Total units
Diesel		
501 to 5324	Built in 1956 to 1958	4,824
Gas		
1044 to 5040	Built in 1956 to 1958	3,997
LPG		
1273 to 5433	Built in 1956 to 1958	4,161

TRACTOR MODEL Farmall 656 / International 656

Gear Drive & Hydro

PLACE OF MANUFACTURE Farmall

RATED HORSEPOWER   PTO   63

NEBRASKA TEST #   909 Gas / 967 Diesel Hydro /
912 Diesel / 968 Gas Hydro

ENGINE SIZE   C263 Gas & LPG   D282 Diesel

BORE & STROKE 3-9/16 x 4.39" Gas & LPG
3-11/16 x 4.39" Diesel

AVERAGE SHIPPING WEIGHT   6,600lbs

TOTAL UNITS BUILT   Farmall   41,404

International   8,741

TRACTOR SERIAL NUMBER DATA

Serial Numbers;	Year built;	Total units
FARMALL		
8501 to 15504	Built in 1965	7,004
15505 to 24373	Built in 1966	8,869
24374 to 32006	Built in 1967	7,632
32007 to 38860	Built in 1968	6,854
38861 to 42517	Built in 1969	3,657
42518 to 45496	Built in 1970	2,979
45497 to 47950	Built in 1971	2,454
47951 to 49904	Built in 1972	1,953
INTERNATIONAL		
7501 to 7841	Built in 1966	341
7842 to 9928	Built in 1967	2,087
9929 11801	Built in 1968	1,874
11802 to 13352	Built in 1969	1,551
13353 to 14193	Built in 1970	841
14194 to 14951	Built in 1971	758
14952 to 15745	Built in 1972	794
15746 to 16241	Built in 1973	495

The 656 was IH's first Hydrostatic drive tractor.

1970 Year built tractors COULD BE an IH Golden Demonstrator

First F-656 built May-12-1965.

TRACTOR MODEL International 660
PLACE OF MANUFACTURE Farmall
RATED HORSEPOWER BELT 81
NEBRASKA TEST # 715 Diesel 721 Gas 722 LPG
ENGINE SIZE D282 Diesel C263 Gas & LPG
BORE & STROKE 3-11/16 x 4-25/64 Diesel
3-9/16 x 4.39" Gas & LPG
AVERAGE SHIPPING WEIGHT 7,900 lbs
TOTAL UNITS BUILT 6,945

TRACTOR SERIAL NUMBER DATA

Serial Numbers;	Year built;	Total units
501 to 3397	Built in 1959	2,897
3398 to 4258	Built in 1960	861
4259 to 5852	Built in 1961	1,594
5853 to 6994	Built in 1962	1,142
6995 to 7445	Built in 1963	450

501	First I-660 Gas
502	First I-660 Diesel
564	First I-660 LPG

TRACTOR MODEL   664
PLACE OF MANUFACTURE Farmall
RATED HORSEPOWER   PTO   54
NEBRASKA TEST #   386
ENGINE SIZE   D-239
BORE & STROKE   3 7/8" x 5.060"
AVERAGE SHIPPING WEIGHT   5,950lbs
TOTAL UNITS BUILT   Unknown

TRACTOR SERIAL NUMBER DATA

Serial Numbers;	Year built;	Total units
2501 to 3211	Built in 1972	711
3212 to --	Built in 1973	

TRACTOR MODEL   International 666
PLACE OF MANUFACTURE Farmall
RATED HORSEPOWER   PTO   66
NEBRASKA TEST #1151 and 1154 Gas
                #1152 and 1155 Diesel
ENGINE SIZE   C291 Gas   D312 Diesel
BORE & STROKE 3.75 x 4.39 Gas   3.875 x 4.410 Diesel
AVERAGE SHIPPING WEIGHT   7,200lbs
TOTAL UNITS BUILT   9,367

TRACTOR SERIAL NUMBER DATA

Serial Numbers;	Year built;	Total units
7501 to 8199	Built in 1972	699
8200 to 11584	Built in 1973	3,385
11585 to 13130	Built in 1974	1,546
13131 to 15738	Built in 1975	2,608
15739 to 16867	Built in 1960	1,128

TRACTOR MODEL   International 686
PLACE OF MANUFACTURE Farmall
RATED HORSEPOWER   PTO   66
NEBRASKA TEST #   1336
ENGINE SIZE   D310 Diesel
BORE & STROKE   3.875 x 4.375
AVERAGE SHIPPING WEIGHT   7,800lbs
TOTAL UNITS BUILT   6,229

TRACTOR SERIAL NUMBER DATA

Serial Numbers;	Year built;	Total units
7500 to 7728	Built in 1976	229
7729 to 9898	Built in 1977	2,170
9899 to 11416	Built in 1978	1,518
11417 to 12922	Built in 1979	1,506
12923 to 13729	Built in 1980	806

First F 686 built   November -2-1976
Last F 686 built   August-19-1980

TRACTOR MODEL   Farmall 706
PLACE OF MANUFACTURE Farmall
RATED HORSEPOWER   PTO   73
NEBRASKA TEST # Gas & LPG 858, 957
                   LPG 860, 956   Diesel 856, 955
ENGINE SIZE   Gas & LPG C263 & C291
                   Diesel D282 & D310
BORE & STROKE   3-9.16 x 4-25/64 (C263)
                   3-11/16 x 4-25/64 (D282)
                   3.75 x 4.39 (C291)
                   3.875 x 4.375 (D310)
AVERAGE SHIPPING WEIGHT   7,800lbs
TOTAL UNITS BUILT   46,147

TRACTOR SERIAL NUMBER DATA

Serial Numbers;	Year built;	Total units
501 to 7072	Built in 1963	6,572
7073 to 21161	Built in 1964	14,089
21162 to 30287	Built in 1965	9,126
30288 to 38520	Built in 1966	8,233
38521 to 46647	Built in 1967	8,126

	First F-706 built June-3-1963
501	First F-706 Gas
502	First F-706 Diesel
510	First F-706 LPG
652	First F-706 Diesel HiClear
7583	First F-706 Diesel All Wheel Drive
37237	First 706 w/D-310 Engine, October-28-1966

TRACTOR MODEL   International 706

PLACE OF MANUFACTURE Farmall

RATED HORSEPOWER   PTO   73

NEBRASKA TEST # Gas 858,957 / LPG 860, 956 /
                 Diesel 856, 955

ENGINE SIZE   Gas & LPG C263 & C291
                 Diesel D282 & D310

BORE & STROKE   3-9.16 x 4-25/64 (C263)
                 3-11/16 x 4-25/64 (D282)
                 3.75 x 4.39 (C291)
                 3.875 x 4.375 (D310)

AVERAGE SHIPPING WEIGHT   8,400lbs

TOTAL UNITS BUILT   5,488

TRACTOR SERIAL NUMBER DATA

Serial Numbers;	*Year built;*	*Total units*
501 to 1250	Built in 1963	750
1251 to 3477	Built in 1964	2,227
3478 to 4788	Built in 1965	1,311
4789 to 5315	Built in 1966	527
5316 to 5988	Built in 1967	672

	First I-706 built	June-3-1963
501	First I-706 Gas	
502	First I-706 Diesel	
588	First I-706 LPG	
1301	First I-706 Diesel All Wheel Drive	
5274	First I-706 with D-310 Engine	

TRACTOR MODEL   Farmall 756
PLACE OF MANUFACTURE Farmall
RATED HORSEPOWER   PTO   76
NEBRASKA TEST #   None
ENGINE SIZE   C291 Gas & LPG D310 Diesel
BORE & STROKE 3.75 x 4.39 (Gas, LPG)
                3.875 x 4.375 (Diesel)
AVERAGE SHIPPING WEIGHT   8,400lbs
TOTAL UNITS BUILT   10,977

TRACTOR SERIAL NUMBER DATA

Serial Numbers;	Year built;	Total units
7501 to 9939	Built in 1967	2,439
9940 to 14124	Built in 1968	4,185
14125 to 17831	Built in 1969	3,707
17832 to 18373	Built in 1970	542
18374 to 18477	Built in 1971	103

First F-756 built   October-3-1967
1970 Year built tractors COULD BE an IH Golden Demonstrator

TRACTOR MODEL   International 756
PLACE OF MANUFACTURE Farmall
RATED HORSEPOWER   PTO   76
NEBRASKA TEST #   None
ENGINE SIZE   C291 Gas & LPG   D310 Diesel
BORE & STROKE 3.75 x 4.39 (Gas, LPG)
                   3.875 x 4.375 (Diesel)
AVERAGE SHIPPING WEIGHT   8,700 lbs
TOTAL UNITS BUILT   927

TRACTOR SERIAL NUMBER DATA

Serial Numbers;	Year built;	Total units
7501 to 7671	Built in 1967	170
7672 to 8163	Built in 1968	492
8164 to 8423	Built in 1969	260
8424 to 8426	Built in 1970	3
8427 to --	Built in 1971	1

First I-756 built   October-4-1967

TRACTOR MODEL   766
PLACE OF MANUFACTURE Farmall
RATED HORSEPOWER   PTO   79
NEBRASKA TEST #1094 Gasoline   1117 Diesel
ENGINE SIZE   C291   D360
BORE & STROKE 3.875 x 4.39 Gas 3.397 x 5.51 Diesel
AVERAGE SHIPPING WEIGHT   9,500 lbs
TOTAL UNITS BUILT   10,800

TRACTOR SERIAL NUMBER DATA

Serial Numbers;	Year built;	Total units
7101 to 7415	Built in 1971	315
7416 to 9610	Built in 1972	2,195
9611 to 12377	Built in 1973	2,767
12378 to 14359	Built in 1974	1,982
14360 to 16839	Built in 1975	2,030
16840 to 17900	Built in 1976	1,060

16541   Black Stripe 766 Tractors start

TRACTOR MODEL 786
PLACE OF MANUFACTURE Farmall
RATED HORSEPOWER PTO 80
NEBRASKA TEST # 1388
ENGINE SIZE D358
BORE & STROKE 3.875 x 5.062
AVERAGE SHIPPING WEIGHT 10,000lbs
TOTAL UNITS BUILT 1,844

TRACTOR SERIAL NUMBER DATA

Serial Numbers;	Year built;	Total units
8601 to 8935	Built in 1980	335
8936 to 10445	Built in 1981	1,509

First 786 built	October-28-1980
Last 786 built	July-2-1981

786 Tractors were fitted with a
4-post ROPS/Canopy OR open platform only.
Open center hydraulics only

TRACTOR MODEL   Farmall 806
PLACE OF MANUFACTURE Farmall
RATED HORSEPOWER   PTO   94
NEBRASKA TEST # 857 Diesel / 859 Gas / 861 LPG
ENGINE SIZE   D361 Diesel   C301 Gasoline & LPG
BORE & STROKE 4-1/8 x 4.5 Diesel
                             3.812 x 4.390 Gas & LPG
AVERAGE SHIPPING WEIGHT   8,800lbs
TOTAL UNITS BUILT   42,958

TRACTOR SERIAL NUMBER DATA

Serial Numbers;	Year built;	Total units
501 to 4708	Built in 1963	4,208
4709 to 15945	Built in 1964	11,237
15946 to 24037	Built in 1965	8,091
24038 to 34942	Built in 1966	10,905
34943 to 43458	Built in 1967	8,515

	First F-806 built	June-12-1963
501	First F-806 Gas	
502	First F-806 Diesel	
531	First F-806 LPG	
526	First F-806 Gas HiClear	
5038	First F-806 Diesel All Wheel Drive	

The 4 Millionth IH Tractor made was a Farmall 806.

TRACTOR MODEL   806 International
PLACE OF MANUFACTURE Farmall
RATED HORSEPOWER   PTO   94
NEBRASKA TEST #857 Diesel / 859 Gas / 861 LPG
ENGINE SIZE   D361 Diesel   C301 Gasoline & LPG
BORE & STROKE   4-1/8 x 4.5 Diesel
                         3.812 x 4.390 Gas & LPG
AVERAGE SHIPPING WEIGHT   9,500 lbs
TOTAL UNITS BUILT   8,053

TRACTOR SERIAL NUMBER DATA

Serial Numbers;	Year built;	Total units
501 to 1402	Built in 1963	902
1403 to 3757	Built in 1964	2,355
3758 to 5916	Built in 1965	2,159
5917 to 7408	Built in 1966	1,492
7409 to 8553	Built in 1967	1,144

	First I-806 built	June-12-63
506	First I-806 Gas	
501	First I-806 Diesel	
504	First I-806 LPG	
1459	First I-806 Diesel All Wheel Drive	

TRACTOR MODEL 826 Farmall and 826 Hydrostatic
PLACE OF MANUFACTURE Farmall
RATED HORSEPOWER   PTO   91
NEBRASKA TEST #   1045   1046 Hydrostatic
ENGINE SIZE D358 Diesel C301 Gasoline & LPG
BORE & STROKE 3.875 x 5.062 Diesel
                    3.812 x 4.390 Gas
AVERAGE SHIPPING WEIGHT   8,500lbs
TOTAL UNITS BUILT   9,589
TRACTOR SERIAL NUMBER DATA

Serial Numbers;	Year built;	Total units
7501 to 8152	Built in 1969	652
8153 to 16351	Built in 1970	8,199
16352 to 17089	Built in 1971	737

1970 Year built tractors COULD BE an IH Golden
Demonstrator

TRACTOR MODEL　　　　　826 International
PLACE OF MANUFACTURE Farmall
RATED HORSEPOWER　　PTO　　　91
NEBRASKA TEST #　　　　1045　　1046 Hydrostatic
ENGINE SIZE　　D358 Diesel　　　C301 Gasoline & LPG
BORE & STROKE　　　　　3.875 x 5.0062 Diesel
　　　　　　　　　　　　　　3.812 x 4.390 Gas
AVERAGE SHIPPING WEIGHT　　　　8,400lbs
TOTAL UNITS BUILT　　　　　　　326

TRACTOR SERIAL NUMBER DATA

Serial Numbers;	*Year built;*	*Total units*
7501 to 7517	Built in 1969	17
7518 to 7718	Built in 1970	200
7719 to 7827	Built in 1971	108

1970 Year built tractors COULD BE an IH Golden
Demonstrator

TRACTOR MODEL   856 Farmall
PLACE OF MANUFACTURE Farmall
RATED HORSEPOWER   PTO   100
NEBRASKA TEST #   970
ENGINE SIZE   D407 Diesel   C301 Gasoline & LPG
BORE & STROKE 4.321 x 4.625 Diesel
                   3.812 x 4.390 Gasoline
AVERAGE SHIPPING WEIGHT   9,700lbs
TOTAL UNITS BUILT   26,894

TRACTOR SERIAL NUMBER DATA

Serial Numbers;	Year built;	Total units
7501 to 9853	Built in 1967	2,353
9854 to 19553	Built in 1968	9,700
19554 to 28692	Built in 1970	9,139
28693 to 32419	Built in 1971	3,727
32420 to 34394	Built in 1972	1,974

First F-856 built   October-3-1967

1970 Year built tractors COULD BE an IH Golden
Demonstrator

TRACTOR MODEL   856 International

PLACE OF MANUFACTURE Farmall

RATED HORSEPOWER   PTO   100

NEBRASKA TEST #   970

ENGINE SIZE   D407 Diesel   C301 Gasoline & LPG

BORE & STROKE 4.321 x 4.625 Diesel
                 3.812 x 4.390 Gasoline

AVERAGE SHIPPING WEIGHT   10,000lbs

TOTAL UNITS BUILT   2,216

TRACTOR SERIAL NUMBER DATA

Serial Numbers;	Year built;	Total units
7501 to 7903	Built in 1967	803
7904 to 9015	Built in 1968	1,112
9016 to 9543	Built in 1969	528
9544 to 9652	Built in 1970	109
9653 to 9716	Built in 1971	63

First I-856 built   October-4-1967

1970 Year built tractors COULD BE an IH Golden Demonstrator

TRACTOR MODEL 886
PLACE OF MANUFACTURE Farmall
RATED HORSEPOWER PTO 86 (D360) 100 (D358)
NEBRASKA TEST # 1254 (D360)   1339 (D358)
ENGINE SIZE D360 and D358
BORE & STROKE 3.875 x 5.085 (D360)
                            3.875 x 5.062 (D358)
AVERAGE SHIPPING WEIGHT   10,500lbs
TOTAL UNITS BUILT   9,579

TRACTOR SERIAL NUMBER DATA

Serial Numbers;	Year built;	Total units
8601 to 10009	Built in 1976	1,409
10010 to 12453	Built in 1977	2,444
12454 to 14413	Built in 1978	1,960
14414 to 15984	Built in 1979	1,571
15985 to 17406	Built in 1980	1,422
17407 to 18179	Built in 1981	770

First 886 built   April-14-1976
Last 886 built   July-2-1981
8601 to 16364   Open Center Hydraulics
            No Closed center hydraulics were built
14472   Start of all D-358 engines

IH built a limited number of RED POWER Demo tractors in 1979 and 1980.

TRACTOR MODEL 966 & 966 Hydrostatic
PLACE OF MANUFACTURE Farmall
RATED HORSEPOWER PTO 100   96   91 Hydro
NEBRASKA TEST # 1123 / 1082 / 1095 Hydrostatic
ENGINE SIZE D-414
BORE & STROKE 4.30 x 5.00
AVERAGE SHIPPING WEIGHT 11,500lbs
TOTAL UNITS BUILT 27,053

TRACTOR SERIAL NUMBER DATA

Serial Numbers;	Year built;	Total units
7101 to 11814	Built in 1971	4,714
11815 to 17793	Built in 1972	5,979
17794 to 22525	Built in 1973	4,732
22526 to 28118	Built in 1974	5,593
28119 to 31771	Built in 1975	3,653
31772 to 34153	Built in 1976	2,381

21697   Last 966 Hydro
31263   Black Stripe 966 Tractors start
34153   Last 966 HiClear

TRACTOR MODEL 986
PLACE OF MANUFACTURE Farmall
RATED HORSEPOWER PTO 106
NEBRASKA TEST # 1255
ENGINE SIZE D436
BORE & STROKE 4.30 x 5.00
AVERAGE SHIPPING WEIGHT 10,600lbs
TOTAL UNITS BUILT 21,095

TRACTOR SERIAL NUMBER DATA

Serial Numbers;	Year built;	Total units
8061 to 11144	Built in 1976	2,544
11145 to 15623	Built in 1977	4,479
15624 to 18981	Built in 1978	3,358
18982 to 22696	Built in 1979	3,715
22697 to 25219	Built in 1980	2,523
25220 to 29155	Built in 1981	3,935

First 986 built   April-14-1976
Last 986 built   July-2-1981

8601 to 23304   Open center Hydraulics
24000 to 26567   Closed center Hydraulics
28000 to 29155   Mexican tractors,
                              Open center hydraulics

IH built a limited number of RED POWER Demo tractors
in 1979 and 1980.

TRACTOR MODEL   1026 Hydrostatic

PLACE OF MANUFACTURE Farmall

RATED HORSEPOWER   PTO   112

NEBRASKA TEST #   1047

ENGINE SIZE   DT407

BORE & STROKE   4.31 x 4.625

AVERAGE SHIPPING WEIGHT   10,000lbs

TOTAL UNITS BUILT   2,415

TRACTOR SERIAL NUMBER DATA

Serial Numbers;	Year built;	Total units
7501 to 9706	Built in 1970	2,205
9707 to 9915	Built in 1971	208

1970 Year built tractors COULD BE an IH Golden Demonstrator

TRACTOR MODEL   1026 International Hydrostatic
PLACE OF MANUFACTURE Farmall
RATED HORSEPOWER   PTO   112
NEBRASKA TEST #   1047
ENGINE SIZE   DT407
BORE & STROKE   4.31 x 4.625
AVERAGE SHIPPING WEIGHT   10,000lbs
TOTAL UNITS BUILT   59

TRACTOR SERIAL NUMBER DATA

Serial Numbers;	Year built;	Total units
7501 to 7549	Built in 1970	49
7550 to 7559	Built in 1971	10

1970 Year built tractors COULD BE an IH Golden
Demonstrator

TRACTOR MODEL   1066 & 1066 Hydrostatic
PLACE OF MANUFACTURE Farmall
RATED HORSEPOWER   PTO   116 & 125
NEBRASKA TEST #   1081 & 1124
ENGINE SIZE   DT414
BORE & STROKE   4.30 x 4.75
AVERAGE SHIPPING WEIGHT   11,800lbs
TOTAL UNITS BUILT   54,948

TRACTOR SERIAL NUMBER DATA

Serial Numbers;	Year built;	Total units
7101 to 12676	Built in 1971	5,576
12677 to 24204	Built in 1972	11,528
24205 to 34848	Built in 1973	10,644
34849 to 46854	Built in 1974	12,006
46855 to 56671	Built in 1975	9,817
56672 to 62048	Built in 1976	5,376

55425   Black Stripe 1066 Tractors start

The 5 Millionth IH built tractor was a 1066,
        made February-1-1974

TRACTOR MODEL   1086
PLACE OF MANUFACTURE Farmall
RATED HORSEPOWER   PTO   131
NEBRASKA TEST #   1247
ENGINE SIZE   DT 414
BORE & STROKE   4.30 x 4.75
AVERAGE SHIPPING WEIGHT   11,700lbs
TOTAL UNITS BUILT   45,745

TRACTOR SERIAL NUMBER DATA

Serial Numbers;	Year built;	Total units
8601 to 14724	Built in 1976	5,824
14725 to 25672	Built in 1977	10,948
25673 to 33846	Built in 1978	8,174
33847 to 42186	Built in 1979	8,340
42186 to 54346	Built in 1980	12,160

First 1086 built   April-9-1976
                   (First of the new 86 series)
Last 1086 built July-2-1981

8601 to 45384   Open center hydraulics
48000 to 54346   Closed center hydraulics
55000 to 56021   Mexican tractors,
                   open center hydraulics

IH built a limited number of RED POWER Demo tractors
in 1979 and 1980.

TRACTOR MODEL　　　　　1206 Farmall
PLACE OF MANUFACTURE Farmall
RATED HORSEPOWER　　PTO　　112
NEBRASKA TEST #　　　910
ENGINE SIZE　　　　　DT 361
BORE & STROKE　　　　4-1/8 x 4.5
AVERAGE SHIPPING WEIGHT　　　　9,400lbs
TOTAL UNITS BUILT　　　　　　　8,403

TRACTOR SERIAL NUMBER DATA

Serial Numbers;	Year built;	Total units
7501 to 8625	Built in 1965	1,125
8626 to 12730	Built in 1966	4,105
12731 to 15903	Built in 1967	9,565

First F-1206　　built September-13-1965

TRACTOR MODEL 1206 International
PLACE OF MANUFACTURE Farmall
RATED HORSEPOWER PTO 120
NEBRASKA TEST # 910
ENGINE SIZE DT 361
BORE & STROKE 4-1/8 x 4.5
AVERAGE SHIPPING WEIGHT 11,100lbs
TOTAL UNITS BUILT 1,590

TRACTOR SERIAL NUMBER DATA

Serial Numbers;	Year built;	Total units
7501 to 7771	Built in 1965	271
7772 to 8491	Built in 1966	720
8492 to 9090	Built in 1967	598

First I-1206 built September-15-1965

TRACTOR MODEL 1256 Farmall
PLACE OF MANUFACTURE Farmall
RATED HORSEPOWER PTO 116
NEBRASKA TEST # 971
ENGINE SIZE DT407
BORE & STROKE 4.321x 4.625
AVERAGE SHIPPING WEIGHT 9,500lbs
TOTAL UNITS BUILT 7,126

TRACTOR SERIAL NUMBER DATA

Serial Numbers;	Year built;	Total units
7501 to 8848	Built in 1967	1,348
8849 to 13139	Built in 1968	4,291
13140 to 14626	Built in 1969	1,486

First F-1256   built October-3-1967

TRACTOR MODEL   1256 International
PLACE OF MANUFACTURE Farmall
RATED HORSEPOWER   PTO   116
NEBRASKA TEST #   971
ENGINE SIZE   DT 407
BORE & STROKE   4.321 x 4.625
AVERAGE SHIPPING WEIGHT   10,900lbs
TOTAL UNITS BUILT   1,232

TRACTOR SERIAL NUMBER DATA

Serial Numbers;	Year built;	Total units
7501 to 7702	Built in 1967	201
7703 to 8443	Built in 1968	740
8444 to 8732	Built in 1969	288

First I-1256   built October-4-1967

TRACTOR MODEL   1456 Farmall
PLACE OF MANUFACTURE Farmall
RATED HORSEPOWER   PTO   131
NEBRASKA TEST #   1048
ENGINE SIZE   DT 407
BORE & STROKE   4.321 x 4.625
AVERAGE SHIPPING WEIGHT   10,500lbs
TOTAL UNITS BUILT   5,583

TRACTOR SERIAL NUMBER DATA

Serial Numbers;	Year built;	Total units
10001 to 10404	Built in 1969	404
10405 to 14148	Built in 1970	3,744
14149 to 15583	Built in 1971	1,435

1970 Year built tractors COULD BE an IH Golden
Demonstrator

TRACTOR MODEL	1456 International
PLACE OF MANUFACTURE	Farmall
RATED HORSEPOWER	PTO 131
NEBRASKA TEST #	1048
ENGINE SIZE	DT407
BORE & STROKE	4.321 x 4.625
AVERAGE SHIPPING WEIGHT	12,600lbs
TOTAL UNITS BUILT	295

TRACTOR SERIAL NUMBER DATA

Serial Numbers;	Year built;	Total units
10001 to 10024	Built in 1969	24
10025 to 10248	Built in 1970	223
10249 to 10295	Built in 1971	47

1970 Year built tractors COULD BE an IH Golden Demonstrator

TRACTOR MODEL   1466
PLACE OF MANUFACTURE Farmall
RATED HORSEPOWER   PTO   133 or 145
NEBRASKA TEST #   1480 & 1125
ENGINE SIZE   DT436
BORE & STROKE   4.30 x 5.00
AVERAGE SHIPPING WEIGHT   12,400lbs
TOTAL UNITS BUILT   25,265

TRACTOR SERIAL NUMBER DATA

Serial Numbers;	Year built;	Total units
7101 to 10407	Built in 1971	3,307
10408 to 15532	Built in 1972	5,125
15533 to 19745	Built in 1973	4,213
19746 to 25403	Built in 1974	5,658
25404 to 29515	Built in 1975	4,112
29516 to 32365	Built in 1976	2,850

28897   Black Stripe 1466 Tractors start

TRACTOR MODEL   1468 V-8 Diesel
PLACE OF MANUFACTURE Farmall
RATED HORSEPOWER   PTO   145
NEBRASKA TEST #   1118
ENGINE SIZE   DV 550 V-8 Diesel
BORE & STROKE   4.50 x 4.3125
AVERAGE SHIPPING WEIGHT   11,800lbs
TOTAL UNITS BUILT   2,906

TRACTOR SERIAL NUMBER DATA

Serial Numbers;	Year built;	Total units
7201 to 7238	Built in 1971	38
7239 to 9108	Built in 1972	1,871
9109 to 9669	Built in 1973	562
9670 to 10106	Built in 1974	4,378

TRACTOR MODEL   1486
PLACE OF MANUFACTURE Farmall
RATED HORSEPOWER   PTO   145
NEBRASKA TEST #   1125 Same as 1466IHC
ENGINE SIZE   DT 436 B
BORE & STROKE   4.30 x 5.00
AVERAGE SHIPPING WEIGHT   11,500lbs
TOTAL UNITS BUILT   21,118

TRACTOR SERIAL NUMBER DATA

Serial Numbers;	Year built;	Total units
8601 to 9797	Built in 1976	1,197
9798 to 14850	Built in 1977	5,053
14851 to 18773	Built in 1978	3,923
18774 to 23161	Built in 1979	4,388
23162 to 27425	Built in 1980	4,264
27426 to 29718	Built in 1981	2,293

First 1486 built   April-30-1976
Last 1486 built   July-2-1981

8601 to 24291   Open center hydraulics
26000 to 28791   Closed center hydraulics

IH built a limited number of RED POWER Demo
Tractors in 1979 and 1980.

TRACTOR MODEL   1566
PLACE OF MANUFACTURE Farmall
RATED HORSEPOWER   PTO   160
NEBRASKA TEST #   1174
ENGINE SIZE   DT436
BORE & STROKE   4.30 x 5.00
AVERAGE SHIPPING WEIGHT   12,800lbs
TOTAL UNITS BUILT   7,418

TRACTOR SERIAL NUMBER DATA

Serial Numbers;	Year built;	Total units
7101 to 7836	Built in 1974	736
7837 to 12588	Built in 1975	4,752
12589 to 14518	Built in 1976	1,930

11929   Black Stripe 1566 Tractors start

TRACTOR MODEL 1568 V-8 Diesel
PLACE OF MANUFACTURE Farmall
RATED HORSEPOWER PTO 160
NEBRASKA TEST # 1175
ENGINE SIZE DV 550 V-8 Diesel
BORE & STROKE 4.50 x 4.3125
AVERAGE SHIPPING WEIGHT 12,900lbs
TOTAL UNITS BUILT 863

TRACTOR SERIAL NUMBER DATA

Serial Numbers;	*Year built;*	*Total units*
7201 to 7820	Built in 1974	620
7821 to 7974	Built in 1975	154
7975 to 8063	Built in 1976	89

7945 Black Stripe 1568 Tractors start

TRACTOR MODEL 1586
PLACE OF MANUFACTURE Farmall
RATED HORSEPOWER PTO 161
NEBRASKA TEST # 1248
ENGINE SIZE DT 436
BORE & STROKE 4.30 x 5.00
AVERAGE SHIPPING WEIGHT 12,900lbs
TOTAL UNITS BUILT 13,637

TRACTOR SERIAL NUMBER DATA

Serial Numbers;	Year built;	Total units
8601 to 10651	Built in 1976	2,051
10652 to 14505	Built in 1977	3,854
14506 to 16346	Built in 1978	1,841
16347 to 18450	Built in 1979	2,104
18451 to 21500	Built in 1980	3,050
21501 to 22237	Built in 1981	737

First 1586 built	April-15-1976
Last 1586 built	June-4-1981

8061 to 18887 Open center hydraulics
21000 to 22237 closed center hydraulics

The 2 Millionth Tractor made at the Farmall Plant was a 1586 RED POWER Demonstrator, built December 5, 1978.

IH built a limited number of RED POWER Demo tractors in 1979 and 1980.

TRACTOR MODEL 3088
PLACE OF MANUFACTURE Farmall
RATED HORSEPOWER PTO 81
NEBRASKA TEST # 1551
      Last IH tractor tested at Nebraska
ENGINE SIZE D358
BORE & STROKE 3.875 x 5.062
AVERAGE SHIPPING WEIGHT 10,000lbs
TOTAL UNITS BUILT 1,366

TRACTOR SERIAL NUMBER DATA

Serial Numbers;	Year built;	Total units
501 to 506	Built in 1981	6
507 to 936	Built in 1982	430
937 to 1420	Built in 1983	484
1421 to —	Built in 1984	-
— to 1866	Built in 1985	445

First 3088 built   August-13-1981

TRACTOR MODEL 3288
PLACE OF MANUFACTURE Farmall
RATED HORSEPOWER PTO 90
NEBRASKA TEST # 1438
ENGINE SIZE D358
BORE & STROKE 3.875 x 5.062
AVERAGE SHIPPING WEIGHT 10,800lbs
TOTAL UNITS BUILT 1,163

TRACTOR SERIAL NUMBER DATA

Serial Numbers;	Year built;	Total units
501 to 1062	Built in 1981	562
1063 to 1284	Built in 1982	222
1285 to 1463	Built in 1983	179
1464 to —	Built in 1984	-
— to 1663	Built in 1985	199

First 3288 built   August-6-1981

TRACTOR MODEL   3388 2+2
PLACE OF MANUFACTURE Farmall
RATED HORSEPOWER   PTO   130
NEBRASKA TEST #   1319
ENGINE SIZE   DT436
BORE & STROKE   4.30 x 5.00
AVERAGE SHIPPING WEIGHT   15,400lbs
TOTAL UNITS BUILT   2,147

TRACTOR SERIAL NUMBER DATA

Serial Numbers;	Year built;	Total units
8801 to 8816	Built in 1978	16
8817 to 10037	Built in 1979	1,221
10038 to 10713	Built in 1980	676
10714 to 10947	Built in 1981	234

First 3388 built   September-19-1978

TRACTOR MODEL   3488 Hydrostatic
PLACE OF MANUFACTURE Farmall
RATED HORSEPOWER   PTO   112
NEBRASKA TEST #   1439
ENGINE SIZE   D466
BORE & STROKE   4.30 x 5.35
AVERAGE SHIPPING WEIGHT   11,500lbs
TOTAL UNITS BUILT   501

TRACTOR SERIAL NUMBER DATA

Serial Numbers;	Year built;	Total units
501 to 714	Built in 1981	214
715 to 722	Built in 1982	8
723 to 828	Built in 1983	106
829 to —	Built in 1984	-
— to 1002	Built in 1985	173

First 3488 built   February-4-1981

TRACTOR MODEL   3588 2+2
PLACE OF MANUFACTURE Farmall
RATED HORSEPOWER   PTO   150
NEBRASKA TEST #   1320
ENGINE SIZE   DT 466B
BORE & STROKE   4.30 x 5.35
AVERAGE SHIPPING WEIGHT   16,500lbs
TOTAL UNITS BUILT   5,644

TRACTOR SERIAL NUMBER DATA

Serial Numbers;	Year built;	Total units
8801 to 8843	Built in 1978	434
8844 to 11797	Built in 1979	2,954
11798 to 13560	Built in 1980	1,763
13561 to 14444	Built in 1981	884

First 3588 built   August-25-1978

TRACTOR MODEL 3688
PLACE OF MANUFACTURE Farmall
RATED HORSEPOWER PTO 113
NEBRASKA TEST # 1440
ENGINE SIZE D436 B
BORE & STROKE 4.30 x 5.00
AVERAGE SHIPPING WEIGHT 11,100lbs
TOTAL UNITS BUILT 2,573
TRACTOR SERIAL NUMBER DATA

Serial Numbers;	Year built;	Total units
501 to 1742	Built in 1981	1,242
1743 to 2481	Built in 1982	739
2482 to 2694	Built in 1983	213
2695 to —	Built in 1984	-
— to 3073	Built in 1985	379

First 3688 built   January-29-1981

14 of the 1982 models were Hi-Clears
30 of the 1983 models were Hi-Clears

TRACTOR MODEL 3788 2+2
PLACE OF MANUFACTURE Farmall
RATED HORSEPOWER PTO 170
NEBRASKA TEST # 1377
ENGINE SIZE DT 466 B
BORE & STROKE 4.30 x 5.35
AVERAGE SHIPPING WEIGHT 16,500lbs
TOTAL UNITS BUILT 2,497
TRACTOR SERIAL NUMBER DATA

Serial Numbers;	Year built;	Total units
8801 to 8807	Built in 1979	7
8808 to 10865	Built in 1980	2,058
10866 to 11297	Built in 1981	432

First 3788 built   August-28-1979

TRACTOR MODEL 3988
PLACE OF MANUFACTURE Farmall
RATED HORSEPOWER PTO
NEBRASKA TEST # No Test
ENGINE SIZE
BORE & STROKE
AVERAGE SHIPPING WEIGHT
TOTAL UNITS BUILT

TRACTOR SERIAL NUMBER DATA
Experimental Tractor

TRACTOR MODEL   4100
PLACE OF MANUFACTURE Farmall
RATED HORSEPOWER   116 Drawbar
NEBRASKA TEST #   931
ENGINE SIZE   DT429
BORE & STROKE   4.50 x 4.50
AVERAGE SHIPPING WEIGHT   13,700lbs
TOTAL UNITS BUILT   1,218

TRACTOR SERIAL NUMBER DATA

Serial Numbers;	*Year built;*	*Total units*
8001 to 8722	Built in 1966	722
8723 to 8985	Built in 1967	264
8986 to 9218	Built in 1968	234

TRACTOR MODEL   4156
PLACE OF MANUFACTURE Farmall
RATED HORSEPOWER   PTO   140
NEBRASKA TEST #   No Test
ENGINE SIZE   DT429
BORE & STROKE   4.50 x 4.50
AVERAGE SHIPPING WEIGHT   14,400lbs
TOTAL UNITS BUILT   229

TRACTOR SERIAL NUMBER DATA

Serial Numbers;	Year built;	Total units
9219 to 9364	Built in 1969	146
9365 to 9447	Built in 1970	83

TRACTOR MODEL 4166
PLACE OF MANUFACTURE Farmall
RATED HORSEPOWER PTO 150
NEBRASKA TEST # 1116
ENGINE SIZE DT436
BORE & STROKE 4.30 x 5.00
AVERAGE SHIPPING WEIGHT 16,000lbs
TOTAL UNITS BUILT 2,566

TRACTOR SERIAL NUMBER DATA

Serial Numbers;	Year built;	Total units
10001 to 10768	Built in 1972	768
10769 to 11254	Built in 1973	486
11255 to 11683	Built in 1974	429
11684 to 12199	Built in 1975	516
12200 to 12566	Built in 1975	366

TRACTOR MODEL 4186

PLACE OF MANUFACTURE Farmall

RATED HORSEPOWER PTO 150

NEBRASKA TEST # 116

ENGINE SIZE DT436

BORE & STROKE 4.30 x 5.00

AVERAGE SHIPPING WEIGHT 16,000lbs

TOTAL UNITS BUILT 701

TRACTOR SERIAL NUMBER DATA

Serial Numbers;	Year built;	Total units
18610 to 18696	Built in 1976	87
18697 to 19030	Built in 1977	334
19031 to 19310	Built in 1978	280

TRACTOR MODEL   4300
PLACE OF MANUFACTURE Hough Mfg, Libertyville, IL
RATED HORSEPOWER   203 Drawbar
NEBRASKA TEST #   815
ENGINE SIZE   DT817
BORE & STROKE   5.375 x 6.00
AVERAGE SHIPPING WEIGHT   29,815lbs
TOTAL UNITS BUILT   44

TRACTOR SERIAL NUMBER DATA

Serial Numbers;	Year built;	Total units
87AH1001 to	Built in 1961	
87AH1045 to	Built in 1965	44

TRACTOR MODEL   4366

PLACE OF MANUFACTURE Steiger Tractor of Fargo, ND

RATED HORSEPOWER   PTO   164

NEBRASKA TEST #   1153

ENGINE SIZE   DT466

BORE & STROKE   4.30 x 5.35

AVERAGE SHIPPING WEIGHT   18,800lbs

TOTAL UNITS BUILT   3,167

TRACTOR SERIAL NUMBER DATA

Serial Numbers;	Year built;	Total units
7501 to 7779	Built in 1973	279
7780 to 8615	Built in 1974	836
8616 to 10226	Built in 1975	1,611
10227 to 10667	Built in 1976	440

First 4366 built   April-13-1973

Last 4366 built   June-17-1976

TRACTOR MODEL　　　　4386
PLACE OF MANUFACTURE Steiger Tractor of Fargo, ND
RATED HORSEPOWER　175 Drawbar
NEBRASKA TEST #　　1256
ENGINE SIZE　　　　DTI 466
BORE & STROKE　　　4.30 x 5.35
AVERAGE SHIPPING WEIGHT　　　　　19,400lbs
TOTAL UNITS BUILT　　　　　　　　2,435

TRACTOR SERIAL NUMBER DATA

Serial Numbers;	Year built;	Total units
501 to 706	Built in 1976	206
707 to 1429	Built in 1977	723
1430 to 2032	Built in 1978	603
2033 to 2205	Built in 1979	173
2206 to 2797	Built in 1980	592
2798 to 2935	Built in 1981	138

First 4386 built	June-25-1976
Last 4386 built	August-27-1981

TRACTOR MODEL   4568
PLACE OF MANUFACTURE Steiger Tractor of Fargo, ND
RATED HORSEPOWER   230 Drawbar
NEBRASKA TEST #   1216
ENGINE SIZE   DVT800
BORE & STROKE   5.312 x 4.500
AVERAGE SHIPPING WEIGHT   19,800lbs
TOTAL UNITS BUILT   858

TRACTOR SERIAL NUMBER DATA

Serial Numbers;	Year built;	Total units
8001 to 8367	Built in 1975	367
8368 to 8858	Built in 1976	492

First 4568 built August-19-1975
Last 4568 built August-19-1977

TRACTOR MODEL 4586
PLACE OF MANUFACTURE Steiger Tractor of Fargo, ND
RATED HORSEPOWER 235 Drawbar
NEBRASKA TEST # No Test, Refer to test #1216
ENGINE SIZE DVT800
BORE & STROKE 5.312 x 4.500
AVERAGE SHIPPING WEIGHT 21,600lbs
TOTAL UNITS BUILT 2,528

TRACTOR SERIAL NUMBER DATA

Serial Numbers;	Year built;	Total units
501 to 814	Built in 1976	314
815 to 1339	Built in 1977	525
1340 to 1945	Built in 1978	606
1946 to 2500	Built in 1979	555
2501 to 2852	Built in 1980	352
2853 to 3028	Built in 1981	176

First 4586 built	June-24-1976
Last 4586 built	August-28-1981

TRACTOR MODEL   4786
PLACE OF MANUFACTURE Steiger Tractor of Fargo, ND
RATED HORSEPOWER   258 Drawbar
NEBRASKA TEST #   1317
ENGINE SIZE   DVT 800
BORE & STROKE   5.312 x 4.500
AVERAGE SHIPPING WEIGHT   21,600lbs
TOTAL UNITS BUILT   2,150

TRACTOR SERIAL NUMBER DATA

Serial Numbers;	Year built;	Total units
501 to 688	Built in 1978	188
689 to 2500	Built in 1979	1,812
2501 to 2555	Built in 1980	55
2556 to 2650	Built in 1981	95

First 4786 built   March-23-1978
Last 4786 built   August-28-1981

TRACTOR MODEL   5000
PLACE OF MANUFACTURE Hinsdale Experimental
RATED HORSEPOWER
NEBRASKA TEST #   No Test
ENGINE SIZE
BORE & STROKE
AVERAGE SHIPPING WEIGHT
TOTAL UNITS BUILT

TRACTOR SERIAL NUMBER DATA
Experimental tractor built at Hinsdale

TRACTOR MODEL 5088
PLACE OF MANUFACTURE Farmall
RATED HORSEPOWER PTO 136
NEBRASKA TEST # 1419
ENGINE SIZE DT 436 B
BORE & STROKE 4.30 x 5.00
AVERAGE SHIPPING WEIGHT 13,500lbs
TOTAL UNITS BUILT 8,177

TRACTOR SERIAL NUMBER DATA

Serial Numbers;	Year built;	Total units
501 to 3350	Built in 1981	2,849
3551 to 6014	Built in 1982	2,464
6015 to 7306	Built in 1983	1,292
7307 to —	Built in 1984	-
— to 8677	Built in 1985	1,370

First Pilot Model 5088 built January-21-1981
First Production 5088 built August-6-1981

TRACTOR MODEL   5188
PLACE OF MANUFACTURE Hinsdale Experimental
RATED HORSEPOWER
NEBRASKA TEST #No Test
ENGINE SIZE
BORE & STROKE
AVERAGE SHIPPING WEIGHT
TOTAL UNITS BUILT

TRACTOR SERIAL NUMBER DATA
Experimental Tractor Built at Hinsdale

TRACTOR MODEL 5288
PLACE OF MANUFACTURE Farmall
RATED HORSEPOWER PTO 163
NEBRASKA TEST # 1420
ENGINE SIZE DT 466 B
BORE & STROKE 4.30 x 5.35
AVERAGE SHIPPING WEIGHT 14,000lbs
TOTAL UNITS BUILT 5,906

TRACTOR SERIAL NUMBER DATA

Serial Numbers;	Year built;	Total units
501 to 2291	Built in 1981	1,790
2292 to 4021	Built in 1982	1,730
4086 to 5053	Built in 1983	968
5054 to —	Built in 1984	-
— to 6406	Built in 1985	1,353

First Pilot Model 5288 built January-18-1981
First Production 5288 built August-6-1981

TRACTOR MODEL　　　　　5488
PLACE OF MANUFACTURE Farmall
RATED HORSEPOWER　　PTO　　187
NEBRASKA TEST #　　　1441
ENGINE SIZE　　　　　DTI 466 C
BORE & STROKE　　　　4.30 x 5.35
AVERAGE SHIPPING WEIGHT　　　14,000lbs
TOTAL UNITS BUILT　　　　3,952

TRACTOR SERIAL NUMBER DATA

Serial Numbers;	Year built;	Total units
501 to 522	Built in 1981	22
523 to 2416	Built in 1982	1,894
2417 to 3111	Built in 1983	695
3112 to —	Built in 1984	-
— to 4452	Built in 1985	1,341

First Pilot Model 5488　　built January-19-1981

4452　　Final Tractor built at Farmall Plant May-14-1985

2WD tractors after 2696 and
All Wheel Drive tractors after 2468 are fitted
　　　　with the Bosch inline injection pump

TRACTOR MODEL   6388
PLACE OF MANUFACTURE Farmall
RATED HORSEPOWER   PTO   130
NEBRASKA TEST #   No Test
ENGINE SIZE   DT 436 B
BORE & STROKE   4.30 x 5.00
AVERAGE SHIPPING WEIGHT   16,100lbs
TOTAL UNITS BUILT   273

TRACTOR SERIAL NUMBER DATA

Serial Numbers;	Year built;	Total units
8801 to 8961	Built in 1981	161
8962 to 9058	Built in 1982	97
9059 to 9071	Built in 1983	13
9072 to 9073	Built in 1984	2

First Production 6388   built August-11-1981

TRACTOR MODEL   6588
PLACE OF MANUFACTURE Farmall
RATED HORSEPOWER   PTO   150
NEBRASKA TEST #   No Test
ENGINE SIZE   DT466 B
BORE & STROKE   4.30 x 5.35
AVERAGE SHIPPING WEIGHT   16,400lbs
TOTAL UNITS BUILT   643

TRACTOR SERIAL NUMBER DATA

Serial Numbers;	Year built;	Total units
8801 to 8965	Built in 1981	164
8966 to 9160	Built in 1982	195
9161 to 9441	Built in 1983	281
9442 to 9443	Built in 1984	2

First Pilot Model 6588   built April-15-1981
First Production 6388   built August-10-1981

TRACTOR MODEL   6788
PLACE OF MANUFACTURE Farmall
RATED HORSEPOWER   PTO   170
NEBRASKA TEST #   1377 Same as 3788
ENGINE SIZE   DT 466 B
BORE & STROKE   4.30 x 5.35
AVERAGE SHIPPING WEIGHT   16,400lbs
TOTAL UNITS BUILT   348

TRACTOR SERIAL NUMBER DATA

Serial Numbers;	Year built;	Total units
8801 to 8809	Built in 1981	9
8810 to 8810	Built in 1982	1
8811 to 9147	Built in 1983	337
9147 to 9148	Built in 1984	2

First Pilot Model 6788   built April-14-1981

TRACTOR MODEL 7288
PLACE OF MANUFACTURE Farmall
RATED HORSEPOWER PTO 175
NEBRASKA TEST # No Test
ENGINE SIZE DTI 466 C
BORE & STROKE 4.30 x 5.35
AVERAGE SHIPPING WEIGHT 19,800lbs
TOTAL UNITS BUILT 19

TRACTOR SERIAL NUMBER DATA

Serial Numbers;	Year built;	Total units
8801 to —	Built in 1984	--
— to 8819	Built in 1985	19

TRACTOR MODEL　　　　　7388
PLACE OF MANUFACTURE Hinsdale Experimental
RATED HORSEPOWER　　　230 ENGINE
NEBRASKA TEST #　　　　No Test
ENGINE SIZE　　　　　　DTI 466 B
BORE & STROKE　　　　　4.30 x 5.35
AVERAGE SHIPPING WEIGHT　　　　　19,400lbs
TOTAL UNITS BUILT　　　　　　　　　3

TRACTOR SERIAL NUMBER DATA
Cancelled prior to full production

Serial Numbers;	Year built;	Total units
501 to 503	Built in 1981	3

First Pilot Model 7388　　built October-5-1981

TRACTOR MODEL 7488
PLACE OF MANUFACTURE Farmall
RATED HORSEPOWER PTO 200
NEBRASKA TEST # No Test
ENGINE SIZE DTI 466 C
BORE & STROKE 4.30 x 5.35
AVERAGE SHIPPING WEIGHT 20,100lbs
TOTAL UNITS BUILT 16

TRACTOR SERIAL NUMBER DATA

Serial Numbers;	Year built;	Total units
8801 to —	Built in 1984	-
— to 8816	Built in 1985	16

TRACTOR MODEL	7588
PLACE OF MANUFACTURE	Hinsdale Experimental
RATED HORSEPOWER	280 ENGINE
NEBRASKA TEST #	No Test
ENGINE SIZE	DVT 800 Diesel
BORE & STROKE	5.312 x 4.500
AVERAGE SHIPPING WEIGHT	21,600lbs
TOTAL UNITS BUILT	2

TRACTOR SERIAL NUMBER DATA
Cancelled prior to full Production

Serial Numbers;	*Year built;*	*Total units*
501 to 502	Built in 1981	2

First Pilot Model 7588   built September-30-1981

TRACTOR MODEL   7688
PLACE OF MANUFACTURE Hinsdale Experimental
RATED HORSEPOWER   PTO   225
NEBRASKA TEST #   No Test
ENGINE SIZE   DTI 466 C
BORE & STROKE   4.30 x 5.35
AVERAGE SHIPPING WEIGHT   20,100lbs
TOTAL UNITS BUILT   1

TRACTOR SERIAL NUMBER DATA

Prototype Tractor at least 1 made -

                              re-decaled & sold as a 7488

TRACTOR MODEL   7788

PLACE OF MANUFACTURE Hinsdale Experimental

RATED HORSEPOWER   300 ENGINE

NEBRASKA TEST #   No Test

ENGINE SIZE   DVT 800 Diesel

BORE & STROKE   5.312 x 4.500

AVERAGE SHIPPING WEIGHT   21,600lbs

TOTAL UNITS BUILT

TRACTOR SERIAL NUMBER DATA

Cancelled prior to full Production

Serial Numbers;	*Year built;*	*Total units*
501 to 503	Built in 1981	3

First Pilot Model 7788   built September-30-1981

TRACTOR MODEL Hydro 60
PLACE OF MANUFACTURE Hinsdale Experimental
NEBRASKA TEST #
ENGINE SIZE
BORE & STROKE
AVERAGE SHIPPING WEIGHT
TOTAL UNITS BUILT

TRACTOR SERIAL NUMBER DATA
Experimental Model

TRACTOR MODEL   Hydro 70
PLACE OF MANUFACTURE Farmall
RATED HORSEPOWER   PTO   70
NEBRASKA TEST # 1154 Gas   1155 Diesel
ENGINE SIZE   C290 Gas   D312 Diesel
BORE & STROKE 3.75 x 4.39   3.875 x 4.410
AVERAGE SHIPPING WEIGHT   7,400lbs
TOTAL UNITS BUILT   3,023

TRACTOR SERIAL NUMBER DATA

Serial Numbers;	Year built;	Total units
7501 to 7569	Built in 1973	69
7570 to 8680	Built in 1974	1,292
8681 to 10093	Built in 1975	1,414
10094 to 10523	Built in 1976	431

TRACTOR MODEL   Hydro 86
PLACE OF MANUFACTURE Farmall
RATED HORSEPOWER   PTO   70
NEBRASKA TEST #   1337
ENGINE SIZE   D310
BORE & STROKE   3.875 x 4.375
AVERAGE SHIPPING WEIGHT   8,000lbs
TOTAL UNITS BUILT   1,907

TRACTOR SERIAL NUMBER DATA

Serial Numbers;	Year built;	Total units
7501 to 7607	Built in 1976	107
7608 to 8170	Built in 1977	563
8171 to 8660	Built in 1978	690
8661 to 9113	Built in 1979	453
9114 to 9407	Built in 1980	294

First Hydro 86   built November-2-1976
Last Hydro 86   built August-15-1980

TRACTOR MODEL   Hydro 100
PLACE OF MANUFACTURE Farmall
RATED HORSEPOWER   PTO   104
NEBRASKA TEST #   1158
ENGINE SIZE   D436
BORE & STROKE   4.30 x 5.00
AVERAGE SHIPPING WEIGHT   11,600lbs
TOTAL UNITS BUILT   5,432

TRACTOR SERIAL NUMBER DATA

Serial Numbers;	Year built;	Total units
7501 to 7726	Built in 1973	226
7727 to 10914	Built in 1974	3,188
10915 to 12433	Built in 1975	1,519
12434 to 12932	Built in 1976	499

12235   Black Stripe HYDRO 100 Tractors start

TRACTOR MODEL　　　　　Hydro 186
PLACE OF MANUFACTURE Farmall
RATED HORSEPOWER　　PTO　　105
NEBRASKA TEST #　　　1257
ENGINE SIZE　　　　　D436 Diesel
BORE & STROKE　　　　4.30 x 5.00
AVERAGE SHIPPING WEIGHT　　　11,800lbs
TOTAL UNITS BUILT　　　　　　4,056

TRACTOR SERIAL NUMBER DATA

Serial Numbers;	Year built;	Total units
8601 to 8812	Built in 1976	211
8813 to 9805	Built in 1977	992
9806 to 10625	Built in 1978	819
10626 to 11464	Built in 1979	838
11465 to 12278	Built in 1980	813
12279 to 12656	Built in 1981	377

First Hydro 186　built April-20-1976
Last Hydro 186　built June-19-1981

8601 to 11620　Open center hydraulics
12000 to 12656　Closed center hydraulics

IH built a limited number of RED POWER Demo
　　　　　　　　　tractors in 1979 and 1980.

IH Casting Codes

Use these letters to translate the letter codes on castings into a year.

Example ...9*26*K means September 26, 1941 or 1964.

1931 -	A	
1932 -	B	
1933 -	C	
1934 -	D	
1935 -	E	
1936 -	F	
1937 -	G	
1938 -	H	
1939 -	I	
1940 -	J	
1941 -	K	
1942 -	L	
1943 -	M	
1944 -	N	
1945 -	O	
1946 -	P	
1947 -	Q	
1948 -	R	
1949 -	S	
1950 -	T	
1951 -	U & W	V was skipped
1952 -	X	
1953 -	Y	
1954 -	Z	
1955 -	A -	The Casting Date letters start over in 1955 and 1977.

IH CHASSIS ALPHABET
USA Built Machines Only   (1939 and later)

A   Distillate
B   Kerosene
C   L.P.Gas
D   5000 ft Altitude
E   8000 ft Altitude
F   Cotton Picker Mounting Attachment (High Drum)
G   Cotton Picker Mounting Attachment (Low Drum)
H   Rear Frame Cover and Shifter Attachment
I   Rockford Clutch
J   Rockford Clutch
K   Optional 4th Gear
L   High Altitude Cylinder Head
M   Low Speed Attachment
N   L.P. Gas Burning Attachment (2500 ft and Above)
O
P   I PTO Attachment without T/A
Q
R   T/A with Provision for Transmission Driven PTO
S   T/A with Provision for 540 I PTO
T   Cotton Picker Mounting Attachment (Low Drum)
U   High Altitude Attachment
V
W   Forward and Reverse Drive
X   High Speed Low and Reverse Attachment
Y   Hydra-Touch Power Supply (w/ 12 GPM Pump)
Z   Hydra-Touch Power Supply (w/ 17 GPM Pump)

IH CHASSIS ALPHABET
USA Built Machines Only (1939 and Later)

AA   1000 RPM I-PTO Drive
BB
CC   3rd Speed Heavy Duty Tillage Gear
DD   High Speed 3rd Gear
EE
FF   Hydraulic Power Supply w/ 4.5 GPM Pump
GG   Hydraulic Power Supply w/ 7 GPM Pump
HH   No Provision for PTO
II
JJ
KK   Optional 4th Gear
LL   Increased Speed Low Gear Attachment
MM
NN   Hydraulic Power Supply
                 (Farmall Tractor only w/ 17 & 9 GPM)
OO
PP   High Capacity Hydraulic Power Supply
                 (International only)

IH ENGINE SUFFIX ALPHABET
USA Built Machines Only (1939 AND LATER)

A	Distillate Burning Attachment
B	Kerosene Burning Attachment
C	L.P.Gas Burning Attachment
D	5000 ft Altitude Attachment
E	8000 ft Altitude Attachment
F	
G	
HA	High Altitude Cylinder Head
I	
J	
K	Cast Iron Pistons
L	High Altitude Cylinder Head
M	
N	L.P. Gas Attachment (2500 ft and above)
O	
P	
Q	
R	Exhaust Valve Rotator
S	
T	
U	High Altitude Attachment
V	Exhaust Valve Rotator
W	
X	
Y	
Z	

IH Tractor Serial Number Locations

FARMALL CUB, INTERNATIONAL CUB and CUB LO-BOY:
Serial Number plate is located on the right side of steering gear housing.

FARMALL SUPER A, AV, B, BN and INTERNATIONAL SUPER A and A-1:
Serial Number plate is located on the right side of tool box/seat support.

FARMALL A, AV, B, BN and INTERNATIONAL A:
Serial number plate is located on the left seat support.

McCORMICK SUPER WD-9, WDR-9 and McCORMICK WDR-9, WR-9, and WR-9S:
Serial number plate is located on the fuel tank support.

INTERNATIONAL CUB 154, 185 AND 184 LO-BOY:
Serial number plate is located on top left front of main frame rail.

IH Tractor Serial Number Locations

FARMALL 544, 656, 666, INTERNATIONAL 284, 544, 656, 2656, 686, 786, 886, 986, 1086, 1486, 1586, HYDRO 186, 3088, 3288, 3388, HYDRO 3488, 3588, 3688, 3788, 5088, 5288, 5488, 6388, 6588, 6788, 7288 and 7488:

> Serial number plate is located on the right side of transmission housing.

INTERNATIONAL 484, 584, 684, 784, 884, HYDRO 84:

> Serial number plate is located on left side of transmission housing.

INTERNATIONAL 600 and 650:

> Serial number plate is located on the right side of fuel tank and air cleaner support.

INTERNATIONAL 4186:

> Serial number plate is located on the left side of front frame below the fender.

INTERNATIONAL 4366, 4568, 4386, 4586, 4786:

> Serial number plate is located on left side of frame near top step.

IH Tractor Serial Number Locations

McCORMICK O-6, ODS-6, OS-6, W-6, WD-6, W-9, WD-9 and INTERNATIONAL I-6, ID-6, I-9 and ID-9:
> Serial number plate is located on the left side of clutch compartment.

FARMALL and INTERNATIONAL C, H, HV, M, MD, MDV, 100, 130, 140, 200, 230, 240, 300, 330, 340, 350, 400, 404, 424, 450, 454, 460, 4100, 4156, 504, 560, 660, 706, 756, 766, 806, 826, 856, 966, 1026, 1066, 1206, 1256, 1456, 1466, 1468, 1566, 1568, HYDRO 100, 3400A, 3500A, 2400A, 2400B, 2410B, 2412B, 2500A, 2510B, 2514B, 2404, 2424, 2504, 2544, 2606, 2656, 2706, 2756, 2806, 2826, 21026, 21206, 21256, 21456, 3414 :
> Serial number plate is located on the left side of clutch housing.

FARMALL SUPERS: C, H, HV, M, MV, MD, MDV, M-T/A:
> Serial number plate is on the left side of clutch housing.

McCORMICK SUPER W-4, W-6, W-6T/A, WD-6, WD-6T/A:
> Serial number plate is located on the left side of clutch housing.

IH Tractor Family
Also referred to as Tractor Series
Photos are representation of Family Series.

CUB: Farmall Cub, International Cub, Cub Lo-Boy, Cub 154, Cub 184, Cub 185.
(Grills Change throughout the production years)

Letter Series: A, AV, B, BN, C, H, HV, M, MD, MDV, MV, O4, OS4, O6, OS6, W4, W6, WD6, W9, WR9S.

Super Series: Super A, Super AV, Super AI, Super A1, Super B, Super C, Super H, Super HV, Super M, Super MV, Super MD, Super MDV, Super MTA, Super MV-TA, Super MTAD, Super MV-TAD, Super W4, Super W6, Super W6TA, Super WD9, Super WDR9.

Hundred: 100, 200, 300, 400, 600.

30 Series: 130, 230, 330

40 Series: 140, 240, 340

50 Series: 350, 450, W450, 650

60 Series: 460, 560, 660

04 Series: 404, 504

06 Series: 606, 706, 806, 1206

26 Series: 826, 1026 Hydro

44 Series: 544

56 Series: 656, 756, 856, 1256, 1456

66 Series: 666, 766, 966, 1066, 1466, 1566

68 Series or "The V-8s": 1458, 1568

86 Series: 686, 786, 986, 1086, 1486, 1586, 186 Hydro

3X88 2+2 "Snoopys": 3388, 3588, 3788

6x88 2+2 "Snoopys": 6388, 6588, 6788

3X88 Row Crop: 3088, 3288, 3488 Hydro

5X88 Row Crop: 5088, 5288, 5488

7X88 4WD: 7388, 7588, 7788

Super 70 STS "Snoopy": 7288, 7488, 7688

66 Series 4WD: 4166, 4366

68 Series 4WD: 4568

86 Series 4WD: 4186, 4386, 4586, 4786

4100 Series: 4100, 4156, 4300

HYDRO Series: Hydro 60, Hydro 70, Hydro 86

NOTES:

NOTES:

www.ingramcontent.com/pod-product-compliance
Lightning Source LLC
Chambersburg PA
CBHW071234210326
41597CB00016B/2052